XLII Symposium on Nuclear Physics 2019

Journal of Physics: Conference Series Volume 1308

Cocoyoc, Mexico
7 - 10 January 2019

Editors:

Luis Acosta
Paulina Amador-Valenzuela
Daniel Jose Marin-Lambarri

ISBN: 978-1-5108-9971-1
ISSN: 1742-6588

Printed from e-media with permission by:

Curran Associates, Inc.
57 Morehouse Lane
Red Hook, NY 12571

Some format issues inherent in the e-media version may also appear in this print version.

Institute of Physics
Dirac House, Temple Back
Bristol BS1 6BE UK

Phone: 44 1 17 929 7481
Fax: 44 1 17 920 0979

techtracking@iop.org

Additional copies of this publication are available from:

Curran Associates, Inc.
57 Morehouse Lane
Red Hook, NY 12571 USA
Phone: 845-758-0400
Fax: 845-758-2633
Email: curran@proceedings.com
Web: www.proceedings.com

TABLE OF CONTENTS

IOP Conf. Series: Journal of Physics: Conf. Series **1308** (2019) 011001 doi:10.1088/1742-6596/1308/1/011001

XLII Symposium on Nuclear Physics

Hacienda Cocoyoc, Morelos, Mexico
January 7-10, 2019

Editors:

Luis Acosta
Instituto de Física
Universidad Nacional Autónoma de México
Paulina Amador-Valenzuela
Instituto Nacional de Investigaciones Nucleares
Daniel José Marín-Lámbarri
Instituto de Física
Universidad Nacional Autónoma de México

IOP Conf. Series: Journal of Physics: Conf. Series **1308** (2019) 011001 doi:10.1088/1742-6596/1308/1/011001

International Advisory Committee

Daniel Bardayan	University of Notre Dame, USA
Francesco Cappuzzello	INFN-Laboratori Nazionali del Sud, Italia
Osvaldo Civitarese	Universidad Nacional de La Plata, Argentina
Jerry Draayer	Louisiana State University, USA
Alfredo Galindo-Uribarri	Oak Ridge National Laboratory, USA
María José García Borge	ISOLDE-CERN, Switzerland
Reiner Krücken	TRIUMF, Canada
Jorge López	University of Texas at El Paso, USA
Takashi Nakamura	Tokyo Institute of Technology, Japan
Thomas Papenbrock	University of Tennessee, USA
Jorge Piekarewicz	Florida State University, USA
Stuart Pittel	University of Delaware, USA
Elena Santopinto	Università degli Studi di Genova, Italia
W. Michael Snow	Indiana University, USA
Adam Szczepaniak	Indiana University, USA
Piet Van Isacker	GANIL, France
Michael Wiescher	University of Notre Dame, USA

Organizing Commmittee

Luis Acosta	IF-UNAM
Paulina Amador-Valenzuela	ININ
Daniel José Marín-Lámbarri	IF-UNAM

Sponsors

Instituto de Ciencias Nucleares, UNAM
Instituto de Física, UNAM
Instituto Nacional de Investigaciones Nucleares
División de Física Nuclear de la SMF

IOP Conf. Series: Journal of Physics: Conf. Series **1308** (2019) 011001 doi:10.1088/1742-6596/1308/1/011001

Photograph of the event

IOP Conf. Series: Journal of Physics: Conf. Series **1308** (2019) 011001 doi:10.1088/1742-6596/1308/1/011001

Preface

As every beginning of the year, the 42nd edition of the Symposium on Nuclear Physics was celebrated at the Hacienda Cocoyoc, most precisely, from January 7th to 10th, 2019. The Symposium was early thought as a small meeting designed to bring together some of the leading nuclear scientists around the world. Its most distinctive feature is to provide a forum for specialists in different areas of nuclear physics: both theorists and experimentalists, students, postdocs, senior and junior scientists, in a relaxed and informal environment, providing them with a unique opportunity to exchange ideas and thus, maybe establish new scientific collaborations as well as to upgrade the already existing. After the first meeting in Oaxtepec in 1978, the Symposium was organized every year without interruption, arriving now to its 42nd edition.

The scientific program consisted of 36 invited talks and 15 posters on a wide variety of hot topics in contemporary nuclear physics, ranging from the traditional fields of nuclear structure and reactions to nuclear astrophysics, hadronic physics, relativistic heavy ions, nuclear instrumentation and applied nuclear physics.

At the meeting there were representatives of many large international collaborations to report on the most recent research progress and the most recent developments in laboratories like Notre Dame, Cyclotron Institute Texas A&M and NSCL/MSU (USA); LUNA-LNGS, LNS and LNL from INFN (Italy); the facilities of TRIUMF (Canada), TU-Dresden (Germany) and iThemba Labs (South Africa), as well as recent results obtained in local Mexican facilities like ININ and LEMA.

On the theoretical side, interesting new developments were presented in nuclear dynamics, nuclear structure, Equation of State, neutrino physics, beyond-the-Standard-Model physics, weakly bound nuclei, nuclear clusters, shell model, hadronic physics and QCD.

The high quality of the talks and posters, the expertise of the speakers and the broad spectrum of subjects covered in the meeting, shows that nuclear physics continues being a very active area at the frontier of scientific research which establishes bridges between many different disciplines, putting together theoretical and experimental researchers and having an extension to the instrumentation developments and applications. Is important to mention the large amount of young researchers and students that participated in this Mexican meeting, giving a nice spirit to the Symposium.

IOP Conf. Series: Journal of Physics: Conf. Series **1308** (2019) 011001 doi:10.1088/1742-6596/1308/1/011001

This special volume of Journal of Physics: Conference Series contains 23 contributions that were presented on this Symposium edition. We want to thank the support received from the Nuclear Physics Division of the Mexican Physics Society, the IF-UNAM and the ICN-UNAM.

Finally, we would like to thank all our colleagues for their enthusiastic participation in this 42nd edition of the meeting. In particular, we are grateful to the members of the International Advisory Committee for their invaluable suggestions and to the anonymous referees for their time and dedication in helping us to maintain the high standard of the contributions published in the proceedings of the Symposium.

We want as well to thank to Nicoll De Los Ríos who repeated as logo-designer for the 42nd edition of this international Symposium.

Luis Acosta
Paulina Amador-Valenzuela
Daniel José Marín-Lámbarri

IOP Conf. Series: Journal of Physics: Conf. Series **1308** (2019) 011002 doi:10.1088/1742-6596/1308/1/011002

42nd
SNP'19

IOP Conf. Series: Journal of Physics: Conf. Series **1308** (2019) 011004 doi:10.1088/1742-6596/1308/1/011004

Peer review statement

All papers published in this volume of *Journal of Physics: Conference Series* have been peer reviewed through processes administered by the proceedings Editors. Reviews were conducted by expert referees to the professional and scientific standards expected of a proceedings journal published by IOP Publishing.

IOP Conf. Series: Journal of Physics: Conf. Series **1308** (2019) 012001 doi:10.1088/1742-6596/1308/1/012001

Heavy radioactive and trans-lanthanide elements in the geothermal microecology of the Los Azufres volcanic complex.

Waleed A Abuhani[1,(a)], Nabanita Dasgupta-Schubert[2]*, Luis Manuel Villaseñor Cendejas[1,(b)], Steven A. Alexander[3] and Sheldon Landsberger[4]

[1] Institute of Physics and Mathematics (IFM), University of Michoacán (UMSNH), C.U., Morelia, Mich. 58060, Mexico
[(a)] Current address: Alghad International Colleges for Applied Medical Science, Al Madinah, Saudi Arabia
[(b)] Current address: University of Morelia (UDEM), Morelia, Mich. 58230, Mexico
[2] Dept. of Physics and Mathematics (FCFM), University of Michoacán (UMSNH), C.U., Morelia, Mich. 58060, Mexico
[3] Dept. of Physics, Southwestern University, Georgetown, TX, USA
[4] Nuclear Engineering Teaching Lab., University of Texas, Austin, TX, USA
* nita@ifm.umich.mx; nitaschubert@gmail.com

Abstract. The distribution of the heavy metals $_{80}Hg$, $_{82}Pb$, $_{90}Th$ and $_{92}U$ in the biogeochemical microcosm of three geo and hydro thermal sites of the Los Azufres volcanic complex have been determined by the nuclear analytical techniques of Polarised Energy Dispersive X-Ray Fluorescence spectrometry and Instrumental Neutron Activation Analysis. We have shown that the sites chosen can be considered to be 'real-time' natural laboratories to simulate the Ordovician-Devonian periods. We observe that (1) the primitive , extremophilic plants of the fumaroles – the mosses and ferns – are efficient agents of pedogenesis; (2) that their rhizospheric soil retains more heavy metal than the hydrothermal sediments. The former implies that large-scale photosynthesis and the consequent burial of soil organic carbon could have substantially contributed to the late-Ordovician climate cooling by CO_2 draw-down, and the latter, that dissolved heavy metals in the hydrothermal seas could have been one of the reasons for the marine extinction of that period. This 'holistic' approach at simulating the Paleozoic environment differs from extant purely laboratory-based studies. However, it corroborates the findings of some earlier works.

1. Introduction and Background Information

The Los Azufres volcanic complex (Fig. 1a) is located 220 km WNW of Mexico City and sits at an altitude of some 2000 m to 3000 m on the central sector of the Mexican volcanic belt. This dormant volcanic complex [1], consists of a large18 km x20 km early Pleistocene caldera and is characterized by many still-active geothermal zones where hot-springs, fumaroles and bubbling mud-pools ("mud volcanoes") abound. The thermal fluids are meteoric water and magmatic fluid in origin with temperatures between 300 °C to 320 °C, high contents of the gases CO_2 and H_2S, pH between 5.8 to

7.2 and an average Cl^- content of 3100 mg/kg. The gas phase may be as much as 90% enriched in CO_2. The average oxygen level is 74% of the value at sea-level and the average yearly precipitation is 14 times higher than the global average. The active geothermal zones form unique microcosms where micro-organisms of the family *archaea* and primitive plants such as mosses (*bryophytes*) and ferns (*pteridophytes*) thrive under these extreme conditions. The sites investigated in this work (Fig. 1b) are Nopalitos-I (N1), Cumbres-I (C1) and Cumbres-II (C2) that are unpopulated, remote and restricted to public thoroughfare. They contain the archetypal fumaroles, hot-springs and hot mud-pools (Fig. 2). Table 1 summarises the physicochemical characteristics of the fumaroles, the hot-springs and the virgin volcanic rock or substrate (VS), also termed "tuff", investigated in this work.

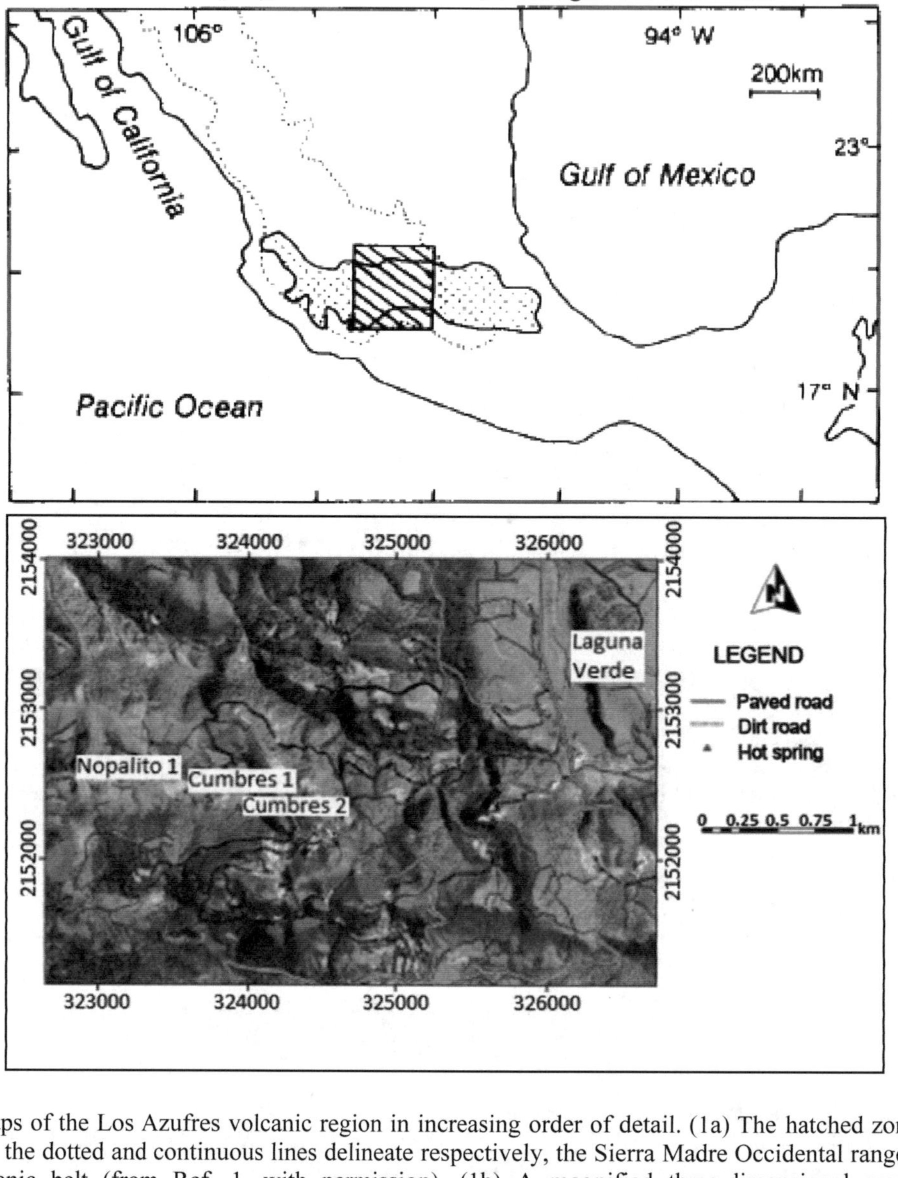

(1a)

(1b)

Fig. 1. The maps of the Los Azufres volcanic region in increasing order of detail. (1a) The hatched zone is Los Azufres, while the dotted and continuous lines delineate respectively, the Sierra Madre Occidental range and the Mexican volcanic belt (from Ref. 1, with permission). (1b) A magnified three-dimensional map of the geothermal zone within Los Azufres showing the sites of Nopalitos-I (N1), Cumbres-I (C1) and Cumbres-II (C2) studied in this work. This figure is a topographical one in the UTM (Universal Transverse Mercator) coordinates and shows the x,y coordinates of the sampled sites (Courtesy Emigdio Casimiro of CFE, Los Azufres, Mexico).

Fig. 2. The bubbling hot mud-pool of N1 and its near-by fumarole showing a wisp of escaping vapour, moss close to its mouth and an overhanging fern.

The first plants to colonise land were the *bryophytes* that appeared in the Ordovician period 485.4 to 443.4 million years ago while the first vascular plants - the fern allies (lycopods) and ferns - appeared in the mid-Devonian period about 360 million years ago [4]. The climate in the early-mid Ordovician was warm and wet and the atmospheric CO_2 and O_2 levels were higher and lower respectively than present day levels. An extensive system of terrestrial and oceanic volcanic arcs spewed forth volcanic gases rich in CO_2, H_2S and SO_2 [5]. The climate in the Devonian period was warm and arid and the atmospheric composition was qualitatively similar to the Ordovician´s. It is believed that these early and ubiquitous land plants were instrumental in pedogenesis (soil formation) through biogeochemical weathering of the bedrock, nutrient recycling and organic matter deposition [6]. A recent work [7] surmises that the Ordovician *bryophytes* were responsible for a large atmospheric CO_2 drawdown through biogeochemical weathering and this led to the late-Ordovician ice-age as well as the eutrophication of the oceans that caused the mass marine extinction of that period. That work however, bases its premise on simulations using a modern non-extremophilic moss grown on rocks in the laboratory and numerical earth system modelling with extrapolated data. Yet another work [8], suggests that the mass marine extinction could have occurred as a consequence of the toxification of the marine organisms by heavy metals such as Fe, Mo, Pb, Mn and As in the paleo sea water.

Table 1: Geo-physicochemical properties of the ambients of the fumaroles and the exposed volcanic substrate (VS) as well as the hot-springs of the sites of N1, C1 and C2 (see ref. 2 for details). The abbreviations RH, GV, NCG stand for relative humidity, geothermal vapour and non-condensible gases, respectively. The GV and NSG values serve as a representative example (well AZ-05 of ref. 3)

Fumaroles and VS:
Air temperature: ~ 45°C (N1), > 45°C (C1, C2), inside mouth; 19°C (N1, C1, C2), outside mouth; 22°C (VS)
RH (%): 69 (N1), 20 (C1, C2), inside mouth; 20 (VS)
Light intensity (kLux): 1.378 (N1) inside mouth; 1.382 x10^2 to 1.117 x10^2 (C2); 1.256x10^2 (VS)
Gas conc. (% of total GV):Steam 97.4, NCG 2.6;
NCG gas conc. (% of total NCG):CO_2 97.4, H_2S 2.3, O_2 0.2, Other gases 0.1

Hot-springs:
Average water temperatures: 80°C (N1), 86.5°C (C1), 88.5°C (C2)
Average pH: 2.31 (N1), 1.95 (C1), 2.00 (C2)
Average electrical conductivity (μS/cm): 3770 (N1), 6580 (C1), 5650 (C2)
Range of concs. (mg/L) of dissolved ions and neutral species: Cl^- 3.2 to 52.7; HCO_3^- 0.0; SiO_2 327 to 456; SO_4^{2-} 843 to 4472.

The micro-ecology of the fumaroles investigated in this work showed the colonization of the mouths of the fumaroles by at least 11 species of mosses and the population of the external rims of the fumaroles by predominantly the ferns (5 distinct species were counted). The physicochemical characteristics (Table 1) point to a close resemblance to the Ordovician and Devonian conditions described earlier. The low light intensity within the fumaroles mimics the cloud cover corresponding to the wet weather during the Ordovician while the high light intensity outside them approximates the lighting condition during the dry warm weather of the Devonian. Taken in sum, the geothermal microcosm studied in this work resembles the planetary ambience of the Ordovician and Devonian periods. This microcosm might then serve as a "real-time" natural laboratory to study the elemental distribution in the bio-geosphere through geothermal activity and biogeochemical weathering by these primitive extremophilic plants. We believe that this approach could possibly better serve as proxy of the environmental processes that might have taken place in the mid-Paleozoic era, than laboratory simulations and model-dependent analyses. The work presented here is an excerpt of our comprehensive study of the elemental distribution within the geothermal microcosm of the sites of N1, C1 and C2 of the Los Azufres volcanic complex. Specifically here, we focus on the distribution of the naturally occurring heavy radioactive elements (NOHRE) ${}_{90}$Th and ${}_{92}$U and the trans-lanthanides (TL), ${}_{80}$Hg and ${}_{82}$Pb, between the rhizospheric soil of the plant, the sediment of the hydrothermal springs and the virgin volcanic substrate also termed "tuff" at the aforesaid sites. It is hoped that this would offer some clues towards understanding how biogeochemical processes versus the purely geochemical ones might have shaped the terrestrial distribution of these elements.

The x-ray radiation-based elemental analysis technique of Polarized Energy Dispersive X-ray Fluorescence (PEDXRF) spectrometry [9] is optimally suited for the multi element analysis entailed in this work because it can analyse elements from Na to U in a single run, thus permitting high sample throughput. It has a large dynamic range (concentrations of the level of percents to parts per million) and it is non-destructive requiring only simple sample preparation. The primary x-ray beam is polarized by double orthogonal reflection and the fluoresced photons are collected in the detector placed at an axis orthogonal to the exciting beam. This 3D optics effectively 'cancels out' the Bremmstrahlung background of the primary beam thereby substantially improving the instrument´s sensitivity. Solid or liquid samples may be analyzed, the only requirement for the former in powder form, being that the particle size should be within 100 μm.

2. Experimental

All details of the experimental procedures involved in this work and the description of the samples are to be found in Ref. 2.

The physical characteristics of density and total organic matter content (TOMC) of the solid substrates were measured after pelletizing the dried, pulverised and sieved substrate powders, under the same uniform high pressure of 10.4 tonnes. The pellet dimensions and weights were recorded. The standard low temperature Loss on Ignition protocol was used to measure the TOMC.

The bulk of the elemental analysis was performed by the SPECTRO XEPOS III (Spectro Analytics GmbH / Ametek Inc., Kleve, Germany) PEDXRF spectrometer at the University of Michoacan (Morelia, Mexico) while a few elements were also analysed by Instrumental Neutron Activation Analysis (INAA) at the University of Texas (Austin, USA). Selected elements of atomic number $Z<13$ were analysed by Inductively Coupled Plasma Mass Spectrometry (ICPMS) at the Helmholtz Centre for Environmental Research (Leipzig, Germany). All the elements reported in this work were analysed by PEDXRF while U was also analysed by INAA. As far as PEDXRF is concerned, in a previous work [10] we have shown that, depending on the matrix and the element analysed, sample masses as low as 0.5g (the factory recommended mass is 4g) may be analyzed without significant deterioration of the statistical figures of merit of the analysis. This permitted us to analyse the low plant dry bio-masses of each type of extremophilic moss collected.

The PEDXRF method used in this work was validated ([2] and [10]) by analysing a series of Standard Reference Materials (SRMs) obtained from the US National Institute of Standards and

Technology (NIST) and SPECTRO Analytical Instruments GmbH (Kleve, Germany). The certificate value for Hg (0.032±0.004) mg/kg for the SRM GSS-1 of SPECTRO was below the detection limit of our spectrometer whereas the measured value for Pb (94.00±1.56) mg/kg showed an index of accuracy of (95.92±6.08)% and a precision factor of 98.35% (*cf.* [10] for the definition of terms). The values for Th and U are given in [2].

The experimental measurements were conducted in replicates of ≥ 3. Errors are quoted as standard deviation (SD) errors on primary measurements and as propagated errors on derived quantities.

3. Results and Discussion

Table 2 shows the average densities and TOMC of the 3 types of substrates that constitute the geothermal microecology: the native volcanic substrate (VS), the hydrothermal spring sediments (S) and the rhizospheric soil (RS) of the extremophilic plants within, or at the rim of, the fumaroles.

Table 2. Average densities and TOMC of the VS, S and RS . The Student´s t-test for statistical distinguishability [11] showed that at 98% Confidence Level, the values for the RS are distinct and the values for the S are not, when compared to the values for the VS [2]

Substrate type	Density (g/cm^3)	TOMC (%)
VS	2.01±0.01	2.18±0.25
S	2.11±0.31	1.78±0.34
RS	1.74±0.18	6.79±1.03

The RS has the least density and the highest amount of TOMC whereas the S has exactly the opposite values. Both the RS and the S however are derived from the same source – the volcanic substrate. Two distinct processes shape them. The RS is formed by biogeochemical weathering of the VS whereas the S is formed by the purely geochemical one. The hot hydrothermal waters rich in highly reactive chemical species (Table 1), break down the volcanic bedrock with the resulting insoluble particles settling as sediment. Rain water run-offs and streams carrying particles resulting from the atmospheric weathering of distant VS also contribute to sediment formation. In contrast, biogeochemical processes are suave and involve the action of root exudates such as organic acids, enzymes and other metabolic products to weather the VS, and the deposition of dead organic matter to form humus. Consequently the substrate in the root zone becomes increasingly soil-like - richer in organic matter and porous. This transition is indicated by an increase in TOMC and a decrease in density as manifested by the RS of this work (Table 2). In terms of the time evolution of the process, we note that the differences of density and TOMC between the native VS and the S are not statistically significant but in the same period the differences between the RS and VS are, which shows that the biogeochemical processes in the rhizosphere of the mosses and ferns are more effective in pedogenesis than the abiotic geochemical ones notwithstanding the harsh chemistry of the hot-spring waters. Hence this microecology that mimics the conditions prevailing during the Ordovician-Devonian periods, shows us that extensive land cover by the first mosses and ferns were instrumental in converting the volcanic rocks to topsoil suitable for the evolution of later higher plants. Furthermore, the three-fold increase of organic matter in the RS compared to the VS means that the fumarolic vegetation is a good sink for atmospheric carbon-dioxide through photosynthesis. Thus the advancing cover of the bare land during the mid-Palaeozoic era by the early ancestors of the same plants studied in this work, was directly (through photosynthesis) and perhaps also largely, responsible for the CO_2 draw-down that could have led to overall climate cooling.

Tables 3a and 3b show the distribution of the 4 heavy metals (HM), Hg, Pb, Th and U, between the VS, S and RS of the geothermal sites of N1 and C2. Elevated concentrations of these metals are highly toxic to all living organism and are known to cause teratological disorders [12]. The VS of the C2 site could not be directly analyzed because it was hard and rocky. Its values in table 3a have been arrived

IOP Conf. Series: Journal of Physics: Conf. Series **1308** (2019) 012001 doi:10.1088/1742-6596/1308/1/012001

at by approximation: the equilibrium ratio of the elemental concentration in the moss RS to its concentration in the corresponding VS was assumed to be equal at the two sites of N1 and C2. Knowing the former at both sites and the latter at the N1 site, the VS (C2) concentration could be arrived at. (The moss RS concentrations are the result of biogeochemical processes that are likely to be the same at the two sites given the same type of overlying vegetation (mosses). Consequently there would exist a definite relationship - linear at its simplest - between the concentration of the element in the substrate with its concentration in the soil formed from the substrate, the RS). The Earth´s crust concentrations were obtained from WebElements (https://www.webelements.com/) which is a compilation from several scholarly works. With the exception of U, the VS concentrations at both sites, but particularly at C2, are distinctly higher than the values for the Earth´s crust which are the averages over a depth of around 60 km and where atmospheric weathering processes do not reach. This observation is a reflection of the high mineral content of the geothermal vapours as a consequence of their magmatic contact, much of which is retained in their condensed phase. Weathering of the VS by acid rain due to the high CO_2 effusions at this geothermal zone, could account for the lowered U concentrations given that uranyl carbonate complexes are soluble, in contrast to the carbonates of Hg, Pb and Th that are not.

Table 3b shows that elemental concentrations in the RS and S at the two sites. It should be mentioned that the RS values for each site shown in the table are the averages over the RS values for the moss and fern RSs at each site (values not shown). For all four elements, RS(C2) accumulates more of the HM than RS(N1). This is not only because the C2 VS is intrinsically richer in these HM (Table 3a) but also because the overlying foliage contained a larger component of ferns and fern allies, whereas the cavity fumaroles of N1 were largely populated by the mosses [*cf.* Ref.2]. The root structure of these first vascular plants are more massive than the small feathery rhizoids of the mosses and therefore are able to interact more with the soil, enhancing biogeochemical activity. Consequently they are better able to fix the HM in the RS. The fixation of the HM occurs through the precipitation of the HM as insoluble organic or inorganic complexes onto soil particles or organic matter and by the adsorption/absorption on the walls of the roots. In fact this fixation in soil is termed *phytostabilization* apropos of the environmental biotechnology of *phytoremediation* which endeavours to remove or contain environmental contamination by plant cover [13]. With the exception of the values for U and the RS(C2) value for Hg, however, the RS values at both N1 and C2, as exemplified by their grand average values, are lower than their corresponding VS values (Table 3a). These fumaroles are wet areas where a substantial amount of the geothermal vapours condense out. The porosity of the RS soil (Table 2) favours soil drainage, that implies a proportionate leach-out of the soluble HM compounds of the RS. The anomalies in the case of Hg and U are possibly due to the degree of insolubility of their organic/inorganic complexes formed by the rhizospheric biogeochemical activity. The physicochemical characteristics of the C2 fumarolic environment differs in detail from that of the N1 fumaroles, which may affect the Hg compounds to a greater extent causing their relatively higher precipitation.

The same observation applies to the case of the S at the two sites. Because only inorganic processes operate in the hydrothermal waters of the hot-springs, and because the gases emanating from the hydrothermal vents and the surface fumaroles close to them have the same origins, it seems plausible that the HM precipitates in both the S and RS are chiefly inorganic in composition. The comparison between the grand averages of the RS and S concentrations throws up an interesting feature - the HM concentrations in the S are generally less than their corresponding ones in the RS, particularly for Hg and Pb (disregarding the statistical variation). This means that leach-outs notwithstanding, the rhizospheric soil is better able to *retain* the HM than the sediments of the hot-springs. Given the high temperatures and high electrical conductivity of the hydrothermal waters and the high concentrations of inorganic anions and colloidal particles (SiO_2) (Table 1), it is likely that a large fraction of the HM remains as soluble or colloidally adsorbed species above the sediments.

Mirroring this evidence back to the Ordovician period, it would mean that the Ordovician´s warm seas would have been substantially richer in the dissolved or dispersed HM species. This would have

IOP Conf. Series: Journal of Physics: Conf. Series **1308** (2019) 012001 doi:10.1088/1742-6596/1308/1/012001

made the HM far more dangerous to aquatic or marine organisms because it is only the solubulized HM that passes through the cell membrane. Thus contrary to the plant cover acting as a the major source of the HM in the run-offs of their leachates to the oceans as stated in [7], it would seem that the prolific terrestrial and underwater volcanic activity of that period could have caused marine extinction by HM toxification, as pointed out in [8].

Table 3a: Average ($n \geq 3$) elemental concentrations of the volcanic substrates VS collected at the sites of N1 and C2, and the average concentration in the Earth´s crust (from WebElements). Z is the element´s atomic number. *Gr.Avg* stands for grand average.
* Derived concentration (see text).

Element	Z	Elemental mass fractions (mg/kg)			
		Earth´s crust	VS(N1)	VS(C2)*	Gr. Avg. VS
Hg	80	0.067	12.85±0.92	14.13	13.49± 0.90
Pb	82	10.00	18.40±3.25	34.11	26.26±11.11
Th	90	6.00	8.35±0.35	17.37	12.86± 6.38
U	92	1.80	0.13±0.002	0.52	0.33± 0.28

Table 3b: Average ($n \geq 3$) elemental concentrations of the rhizospheric soil (RS) and the hot-spring sediment (S) samples collected from the sites of N1 and C2.

Element	RS(N1)	RS(C2)	Gr.Avg. RS	S(N1)	S(C2)	Gr.Avg. S
Hg	9.40±3.68	23.89±22.89	16.64±10.24	5.98±2.16	18.83±3.85	12.40±9.09
Pb	13.18±1.45	28.55±5.69	20.86±10.87	14.13±0.39	21.95±0.85	18.04±5.53
Th	6.08±0.25	11.75±3.55	9.86±4.02	5.95±0.28	12.55±2.05	9.25±3.99
U	0.15±0.02	0.81±0.39	0.60±0.44	0.21±0.02	1.17±1.06	0.69±0.83

4. Conclusion

We have studied the distribution of the heavy metals $_{80}$Hg, $_{82}$Pb, $_{90}$Th and $_{92}$U in the geothermal and hydrothermal biogeochemical microcosm of the Los Azufres volcanic complex and have shown that the sites chosen can be considered to be real-time natural laboratories to proxy the Ordovician-Devonian periods. The heavy metals were traced by Polarised Energy Dispersive X-ray Fluorescence spectrometry and Instrumental Neutron Activation Analysis. While the statistical variation is large as would be obvious given the multitude of variables that are involved in such a non-controlable environment, certain trends could nonetheless be clearly observed. We note the following: (1) The primitive, extremophilic plants of the fumaroles – the mosses and ferns – are efficient agents of pedogenesis; (2) that their rhizospheric soil retains more of the heavy metals than the hydrothermal sediments. The former implies that large-scale photosynthesis and the consequent burial of soil organic carbon could have substantially contributed to the late-Ordovicial climate cooling by CO_2 draw-down, and the latter, that dissolved heavy metals in the hydrothermal seas could have been one of the reasons for the marine extinction of that period. The aforesaid 'holistic' approach at simulating the Paleozoic environment differs from extant purely laboratory based studies. However, it corroborates the findings of some earlier works.

References

[1] Ferrari L *et al* 1991 *Journal of Volcanology and Geothermal Research*, **47**, 129-148
[2] Abuhani WA et al 2015 Journal of Environmental Radioactivity, **139**, 33-42
[3] Suarez Arriaga MC et al 2000 *Proceedings of the World Geothermal Congress, Kyushu Tohoku,*

Japan, May 28-June10
[4] Stewart WN, Rothwell GW. *Paleobotany and the evolution of plants*. 2nd ed. 1993. Cambridge Univ. Press , Cambridge, UK
[5] Parnell J, Foster S. 2012 *Geochemical Transactions*, **13**(1), 7
[6] Jenny H, *Factors of soil formation*. 1994. Dover Publications Inc., New York
[7] Lenton T *et al* 2012 *Nature Geoscience* **5**, 86-89
[8] Vanderbroucke TRA *et al* 2015 *Nature Communications* **6**, 7966
[9] Heckel J, Ryon RW 2002 *Polarized beam x-ray fluorescence analysis*. In: Van Griecken R., Markowicz A. (Eds). Handbook of x-ray spectreometry, 2nd ed. Marcel Dekker Inc. New York, pp. 603-630
[10] Abuhani WA et al 2014 *Powder Diffraction* **29**, 159-169
[11] Mandel J. 1984. The Statistical Analysis of Experimental Data. Dover Publications, New York.
[12] Inouye M 1989 *Congenital Anomalies*, **29**(4), 333-344
[13] Pilon-Smits E 2005 *Annual Review of Plant Biology*, 56, 15-39 (2005).

IOP Conf. Series: Journal of Physics: Conf. Series **1308** (2019) 012002 doi:10.1088/1742-6596/1308/1/012002

Recent results on Heavy-Ion induced reactions of interest for *0νββ* decay

C. Agodi[1], F. Cappuzzello[1,3], L. Acosta[20], C. Altana[1], P. Amador-Valenzuela[21], N. Auerbach[27], J. Barea[34], J.I. Bellone[1], R. Bijker[22], D. Bonanno[2], T. Borello-Lewin[14], I. Boztosun[25], V. Branchina[2,3], S. Brasolin[4], G. Brischetto[1,3,5], O. Brunasso[4], S. Burrello[1], S. Calabrese[1,3], L. Calabretta[1], D. Calvo[4], V. Capirossi[4,6], D. Carbone[1], M. Cavallaro[1], R. Chen[33], L.E. Charon Garcia[20], E .R. Chávez Lomelí[20], I. Ciraldo[1,3], M. Colonna[1], G. D'Agostino[1], H. Djapo[25], G. De Geronimo[32], F. Delaunay[4,6,12], K. De Los Rios[20], N. Deshmukh[30], C. Ferraresi[4,7], J.L. Ferreira[15], J. Ferretti[8,13], P. Finocchiaro[1], S. Firat[25], M. Fisichella[4], A. Foti[2], G. Gallo[1,3], H. Garcia-Tecocoatzi[8,9], A. Hacisalihoglu[26], A. Huerta-Hernandez[20], J. Kotila[31], Y. Kucuk[25], F. Iazzi[4,6], L. La Fauci[1,3], G. Lanzalone[1,10], F. La Via[11], J.A. Lay[28] , H. Lenske[23], R. Linares[15],D. Lo Presti[2,3], J. Lubian[15], J. Ma[33], D. Marín-Lámbarri[20], J. Mas Ruiz[20], N. H. Medina[14], D.R. Mendes[15], P. Mereu[4], M. Moralles[16], L. Neri[1], J. R. B. Oliveira[14], A. Pakou[18], L. Pandola[1], H. Petrascu[29], N. Pietralla[24], F. Pinna[4,6], S. Reito[2], P. Ries[24], M.R.D. Rodrigues[14], A. D. Russo[1], G. Russo[2,3], E. Santopinto[8], R.B.B. Santos[17], L. Serbina[29], O. Sgouros[1], M.A.G. da Silveira[17] , S.O. Solakcı[25], G. Souliotis[19],V. Soukeras[1], A. Spatafora[1,3], D. Torresi[1], S. Tudisco[1], R.I.M. Vsevolodovna[8,9] , H. Vargas Hernandez[20] , J.S. Wang[33], V. Werner[24], Y.Y. Yang[33], A. Yildirim[25], V.A.B. Zagatto[15] for the NUMEN collaboration

[1] *Istituto Nazionale di Fisica Nucleare, Laboratori Nazionali del Sud , Italy*

[2] *Istituto Nazionale di Fisica Nucleare, Sezione di Catania, Italy*

[3] *Dipartimento di Fisica e Astronomia "Ettore Majorana", Università di Catania, Italy*

[4] *Istituto Nazionale di Fisica Nucleare, Sezione di Torino, Italy*

[5] *Centro Siciliano di Fisica Nucleare e Struttura della Materia, Italy*

[6] *DISAT, Politecnico di Torino, Italy*

[7] *DIMEAS, Politecnico di Torino, Italy*

[8] *Istituto Nazionale di Fisica Nucleare, Sezione di Genova, Italy*

[9] *Dipartimento di Fisica, Università di Genova, Italy*

[10] *Università degli Studi di Enna "Kore", Italy*

[11] *CNR-IMM, Sezione di Catania, Italy*

[12] *LPC Caen, Normandie Université, ENSICAEN, UNICAEN, CNRS/INP3, France*

[13] *Department of Physics, Yale University, USA*

[14] Instituto de Fisica, *Universidade de Sao Paulo, Brazil*

[15] Instituto de Fisica, *Universidade Federal Fluminense, Brazil*

[16] *Instituto de Pesquisas Energeticas e Nucleares IPEN/CNEN, Brazil*

[17] *Centro Universitario FEI Sao Bernardo do Campo, Brazil*

[18] Department of Physics and HINP, University of Ioannina, Greece

[19] Department of Chemistry, National and Kapodistrian University of Athens, Greece

[20] Instituto de Fìsica, Universidad Nacional Autónoma de México, Mexico

[21] Instituto Nacional de Investigaciones Nucleares, Mexico

[22] Instituto de Ciencias Nucleares, Universidad Nacional Autónoma de México, Mexico

[23] Department of Physics, University of Giessen, Germany

[24] Institut fur Kernphysik, Technische Universitat Darmstadt, Germany

[25] Department of Physics, Akdeniz University, Turkey

[26] *Institute of Natural Sciences, Karadeniz Teknik University, Turkey*

[27] School of Physics and Astronomy, Tel Aviv University, Israel

[28] Departamento de FAMN, University of Seville, Spain

[29] *Department of Nuclear Physics, Horia Hubulei National Institute of Ph*ysics and Nuclear Engineering, IFIN-HH, Romania

[30] Nuclear Physics Division, Saha Institute of Nuclear Physics, India

[31] University of Jyväskylä, Finland

[32] Stony Brook University, USA

[33] Institute of Modern Physics, Chinese Academy of Sciences, Lanzhou, China

[34] *Departamento de Fìsica, Universidad de Conception, Chile*

agodi@lns.infn.it

Abstract. An updated overview of recent results on Heavy-Ion induced reactions of interest for neutrinoless double beta decay is reported in the framework of the NUMEN project. The NUMEN idea is to study heavy-ion induced Double Charge Exchange (DCE) reactions with the aim to get information on the nuclear matrix elements for neutrinoless double beta (0νββ) decay. Moreover, to infer the neutrino average masses from the possible measurement of the half- life of 0νββ decay, the knowledge of the nuclear matrix elements is a crucial aspect.

IOP Conf. Series: Journal of Physics: Conf. Series **1308** (2019) 012002 doi:10.1088/1742-6596/1308/1/012002

1. Introduction

The search for neutrinoless double beta decay (*0νββ*) addresses fundamental questions in physics, related to the absolute neutrinos mass scale, whether neutrinos are Majorana particles or how many neutrino species are there. Moreover *0νββ* search has implications for the understanding of the matter-antimatter asymmetry in the universe. The design of the *0νββ* search experiments themselves require detailed understanding of nuclear physics, as background critically depend on nuclear decay chain and nuclear matrix elements are needed to link observed rates to neutrino properties.

In this framework the NUMEN project propose an innovative and original tool: to use Heavy-Ions Double Charge Exchange Reactions (HI-DCE) to access quantitative information for Nuclear Matrix Element towards *0ββ* decay.

2. The NUMEN Project

The NUMEN main goal is the extraction from measured cross-sections of "data-driven" information on Nuclear Matrix Elements for all the systems candidate for *0νββ*. Crucial for the experimental challanges is the INFN Laboratori Nazionali del Sud (LNS) facility, made up by the Superconducting Cyclotron (CS) and the MAGNEX magnetic spectrometer. The experimental measurements of HI-DCE reactions present a number of challenging aspects, since they are characterized by very low cross sections and, in the same experimental conditions, they must be identified with respect to competitive reaction channels.

It is well known that the *0νββ* half-life can be factorized in three terms at different physics scale: the phase-space factor, connected with Atomic physics, the Matrix Element (NME) related with Nuclear physics and a term, related to Particle physics, in which it is supposed there are the answers to the unsolved questions, related to new physics beyond the Standard Model. For evaluation of NME several methods have been used, based on different nuclear models, an updated comparison of the results of NME calculations, obtained within various nuclear structure frameworks [1-4], indicates that significant differences are indeed found. In addition, some assumption common to different competing approaches could cause overall systematic uncertainties.

NUMEN [5, 6] propose a novel way to address experimentally-driven information on the NMEs of 0νββ, based on DCE cross section measurements. These reactions are characterized by the transfer of two charge units, leaving the mass number unchanged, and can proceed either by a sequential multi-nucleon-transfer mechanism or by meson-exchange. Despite *0νββ* decays and HI-DCE reactions are mediated by different interactions, they present a number of similarities. The main ones are:

i) the initial and final state wave functions in the two processes are the same,
ii) the transition operators are similar, in both cases Fermi, Gamow-Teller and rank-two tensor components are present,
iii) a large linear momentum (100 MeV/c) is available in the virtual intermediate channel,
iv) the two processes are non-local and are characterized by two vertices localized in a pair of valence nucleons,
v) they take place in the same nuclear medium,
vi) a relevant off-shell propagation through virtual intermediate channels is present.

The descriptions of NMEs for DCE and 0νββ decay present the same degree of complexity, with the advantage for DCE to be "accessible" in laboratory. However, a simple relation between DCE cross sections and 0νββ decay half-lives is not trivial and needs to be explored.

At the MAGNEX facility of the INFN-LNS we demostrate the feasibility of this kind of DCE cross sections measurements. MAGNEX is a large acceptance magnetic system able to provide high resolution in energy, mass and angle [7] and an accurate control of the detection efficiency. The implementation of trajectory reconstruction technique is the key feature of MAGNEX, which guarantees the above mentioned performance and its relevance in the research for heavy-ion physics [8-10], also taking advantage of its coupling to the EDEN neutron detector array [11,12]. The ^{40}Ca(^{18}O,^{18}Ne)^{40}Ar reaction was studied at 15 MeV/u, showing that high mass, angular and energy resolution energy spectra and accurate absolute cross sections are at our reach, even at very forward angles, see ref. [13]. In addition, a schematic analysis of the reaction cross sections demonstrated that relevant quantitative information on DCE matrix elements can be extracted from the data. This result demonstrated that the previous experimental limitations are almost overcome and that high resolution and statistically significant experimental data can be measured for DCE processes.

In the present experimental conditions, due to the limitation arising from the tiny cross-sections of the processes of interest, only very few systems can be measured within the 5-years project. In order to systematically explore all the nuclei candidates for 0νββ, a beam intensity at least two orders of magnitude higher than the present must be achieved.

The NUMEN project is conceived in a long-range time perspective, in view of a comprehensive study of many candidate systems for *0νββ* decay. Moreover, the project promotes a renewal of the INFN-LNS research infrastructure [14] with a specific R&D activity on detectors, materials and instrumentation. As a consequence, major upgrades of the detector technologies (3D ion tracker, particle-identification wall, gamma-ray array, ...) must be developed [15, 20]. Also the target technology must be upgraded, to avoid the damage of the thin films due to the high temperature involved [21, 23]. Moreover, a deep and complete investigation of the theoretical aspects connecting nuclear reaction mechanisms and nuclear matrix elements must be carried out [24, 26].

2.1. *An updated overview*

NUMEN experimental activity consists of two main classes of experiments, corresponding to the exploration of the two directions of isospin transfer τ− τ− and τ+ τ+, characteristic of β−β− and β+β+ decays, respectively. In particular, the β+β+ direction in the target is investigated using a ^{18}O beam and measuring the (^{18}O,^{18}Ne) DCE induced transitions, together with the other reaction channels involving the same beam and target. Similarly, the β−β− direction is explored via the (^{20}Ne,^{20}O) reaction, using a ^{20}Ne beam and detecting the reaction products of the DCE channel along with other open channels characterized by the same projectile and target. With the actual facility and experimental setup we perform some long run at LNS with MAGNEX, choosing few isotopes, candidates for 0νββ, already at our reach in terms of energy resolution and availability of thin targets.

In particular, we performed for the first time experimental investigations of the (^{20}Ne,^{20}O) DCE reaction on ^{116}Cd,^{76}Ge and ^{130}Te targets. These are the first measurements of such a kind of reaction: no data are available in literature. A ^{20}Ne^{10+} cyclotron beam at 15 AMeV was delivered by the CS of INFN-LNS and impinged on ^{116}Cd rolled target of 1370 μg/cm^2 thickness and ^{76}Ge (386 μg/cm^2 thickness) and ^{130}Te (247 μg/cm^2 thickness) both evaporated on a C backing of 50 μg/cm^2. The thickness of the various targets was carefully chosen in order to obtain an energy resolution which allows to distinguish the transition to the residual nucleus ground state from its first excited state. Indeed, the selected thickness of ^{116}Cd is much higher than that of ^{76}Ge and ^{130}Te, because the first excited state in ^{116}Sn case is at 1.293 MeV, to be compared to 0.559 MeV in ^{76}Se and 0.536 MeV in ^{130}Xe. The MAGNEX spectrometer was placed at forward angles including zero degrees in the full acceptance mode (50 msr). Despite the experimental limitations, we were able to measure energy spectra and absolute cross sections for the DCE reaction channel. Moreover, we measured also other reaction channels: one- and two-proton transfer, one- and two-neutron transfer and Single Charge Exchange, in order to estimate the role of the sequential multi-nucleon transfer routes on the diagonal DCE process.

IOP Conf. Series: Journal of Physics: Conf. Series **1308** (2019) 012002 doi:10.1088/1742-6596/1308/1/012002

3. Conclusions and outlook

Pioneering experiments on (^{18}O,^{18}Ne) and (^{20}Ne,^{20}O) DCE performed at INFN-LNS Laboratory in Catania have shown that accurate cross sections measurements at very forward angles can be done. The main activity characterizing the NUMEN project is the measurement of DCE absolute cross sections, with the related processes, and the extraction of relevant NMEs. The NUMEN "Holy Grail" is to find a connection between the NMEs extracted from DCE reactions and those characterizing 0νββ decay, at least in the nuclei in which this process is energetically allowed. NUMEN is a big challenge both for nuclear technology and nuclear theory. Moreover it promotes an important upgrade of the experimental facilities and the development of the different theoretical aspects, connected with the nuclear structure and reaction mechanisms, involved in heavy ions induced in DCE reactions. High intensity beams are the new frontier for these challenging studies. NUMEN indicates a possible growth prospect of heavy-ion physics in synergy with neutrino physics, aims at giving an innovative contribution in one of the most promising fields of fundamental physics.

References

[1] J. D. Vergados et al., Reports on Progress in Physics 75, 106301 (2012).
[2] P. Vogel, Jour. of Phys. G. 39, 124002 (2012).
[3] J. Barea, J. Kotila, F. Iachello, Phys. Rev. Lett. 109, 042501 (2012).
[4] S. Dell'Oro et al., Advances in High Energy Physics 2016, 216259 (2016)
[5] F. Cappuzzello et al., Eur. Phys. J. A 54, 72 (2018).
[6] C. Agodi et al., Nucl. Part. Phys. Proc. 265, 28 (2015).
[7] F. Cappuzzello et al., Eur. Phys. J. A 52, 167 (2015).
[8] F. Cappuzzello et al., Nature Communications 6, 6743 (2015).
[9] D. Carbone et al., Phys. Rev. C 95, 034603 (2017).
[10] J. R. B. Oliveira et al., J. Phys. G 40, 105101 (2013).
[11] M. Cavallaro et al., Phys. Rev. C 93, 064323 (2016).
[12] M. Cavallaro et al., Nucl. Instr. and Meth. A 700, 65 (2013).
[13] F. Cappuzzello et al., Eur. Phys. J. A 51, 145 (2015).
[14] L. Calabretta et al., Modern Physics Letters A 32, 17 (2017).
[15] M. Cortesi et al., Review of Scientific Instruments 88, 013303 (2017).
[16] D. Lo Presti et al., J. Phys. Conf. Ser. 18, 012034 (2018).
[17] S. Tudisco et al., Sensors 18, 2289 (2018).
[18] A. Muoio et al., EPJ Web of Conferences 117, 10006 (2016).
[19] D. Carbone et al., Results in Physics 6, 863-865 (2016).
[20] J. R. B. Oliveira et al., J. Phys. Conf. Ser. 18, 012040 (2018).
[21] F. Pinna et al., J. Phys. Conf. Ser. 1056, 012046 (2018).
[22] F. Iazzi et al., WIT Transactions on Engineering Science 116, 206 (2017).
[23] S. Calabrese et al., Acta Phys. Pol. B 49, 275 (2018).
[24] H. Lenske et al., Phys. Rev. C 98, 044620 (2018).
[25] E. Santopinto et al., Phys. Rev. C 98, 061601(R) (2018).
[26] J. Bellone et al., J. Phys. Conf. Ser. 1056, 012004 (2018).

Scalar current limit from the beta-neutrino correlation: the WISArD experiment

V. Araujo-Escalona[1], P. Alfaurt[2], P. Ascher[2], D. Atanasov[1,3], B. Blank[2], L. Daudin[2], X. Fléchard[4], M. Gerbaux[2], J. Giovinazzo[2], S. Grévy[2], T. Kurtukian Nieto[2], E. Liénard[4], L. Nies[5], G. Quéméner[4], M. Roche[2], N. Severijns[1], S. Vanlangendonck[1], M. Versteegen[2], P. Wagenknecht[1], D. Zakoucky[6].

[1] KU Leuven, Department of Physics and Astronomy, Instituut voor Kern- en Stralingsfysica, Celestijnenlaan 200 D, B-3001 Leuven, Belgium.
[2] CEN Bordeaux-Gradignan, 19 Chemin du Solarium, CS 10120, F-33175 Gradignan, France.
[3] CERN, CH-1211 Genève 23, Switzerland.
[4] LPC Caen, ENSICAEN, Université de Caen, CNRS/IN2P3, Caen, France.
[5] II. Physics Institute, University of Giessen, Germany.
[6] Nuclear Physics Institute CAS, Řež, Czech Republic.

E-mail: victoria.araujoescalona@kuleuven.be

Abstract. Beta - neutrino correlation measurements are key in the research of physics beyond the Standard Model. In pure Fermi beta transitions, the beta-neutrino correlation coefficient, $a_{\beta\nu}$, is sensitive to the presence of scalar currents. The present limits were established by experimental studies of various nuclear systems with allowed Fermi transitions. A new experiment to improve the constraints on scalar currents is being developed, by the WISArD collaboration at ISOLDE/CERN, where the aim is to measure the energy shift of the β-delayed protons emitted from the isobaric analogue state of the ^{32}Ar ground state. To enhance the sensitivity, protons and positrons are guided by a strong magnetic field and measured in coincidence between the two detection configurations located on both sides of a catcher foil in which the radioactive samples are implanted. Kinematic energy shifts of the protons in coincidence with positrons, in the same or opposite hemisphere of the catcher foil, will be more or less pronounced as a function of the possible scalar current component of the weak interaction. Details of the apparatus and preliminary results of the experiment are presented.

1. Introduction

The Vector – Axial – vector (V-A) formalism [1] of nuclear β decay has been established a few decades ago and ever since has been continuously tested in an effort to find physics beyond the Standard Model (SM). There are two main experimental approaches to test the formalism: one is to perform collider experiments [2, 3] at very high energies to find evidence of new particles, another one uses low-energy β-decay experiments to determine different aspects of the interaction [4]. The latter approach is a primary method of new experiments that provide high precision tests of the discrete symmetries in the SM, and search for non-SM interaction components. Experimental error bars still leave sufficient room for the possible existence of other types of weak interaction in β decay, i.e. scalar (S), tensor (T), and pseudoscalar (P) type interactions. Together with the well-known vector (V) and axial -vector (A) interactions, they appear in the most general form of the β decay Hamiltonian at the present level of precision. The P interaction can be neglected in the non-relativistic description of nucleons [5]. The coupling constants C_i and C_i' (i = S, V, A and T) for the different types of weak interaction have to be determined experimentally.

2. Weak interaction studies with radioactive samples

The experimental determination of the magnitude of the coupling constants (C_i and C_i') for a given exotic current is well described by Jackson et al. [6] by using the decay rate distribution. Following their formalism, the decay rate is given as a function of the total energies and momenta of the particles of the decaying system, including their spins. For example, the angular and energy distributions of the electron and neutrino in the case of a decay of unpolarized nuclei or neutrons will have the form

$$\omega(E_e) = \omega_o(E_e)\xi\Big(1 + \frac{\vec{p}_e \cdot \vec{p}_\nu}{E_e E_\nu}a + \frac{m_e}{E_e}b + ...\Big) \tag{1}$$

where the function $\omega_o(E_e)$ represents the phase space factor and the Fermi function, m_e is the rest mass of the electron and E_e is the total energy of the β particle. The coefficients a ($\beta - \nu$ angular correlation) and b (Fierz interference term) are called correlation coefficients. Each coefficient measurement contains specific information on the dynamics of the decay, expressed in terms of Fermi (F) and Gamow-Teller (GT) nuclear matrix elements, M_F and M_{GT} respectively, and the coupling constants C_i and C_i', with ξ defined as

$$\begin{aligned}\xi =& \mid M_F \mid^2 \big(\mid C_S \mid^2 + \mid C_V \mid^2 + \mid C_S' \mid^2 + \mid C_V' \mid^2 \big) + \\ & \mid M_{GT} \mid^2 \big(\mid C_T \mid^2 + \mid C_A \mid^2 + \mid C_T' \mid^2 + \mid C_A' \mid^2 \big)\end{aligned} \tag{2}$$

This work is concentrating on the search of scalar currents. It requires the determination of the $\beta - \nu$ correlation coefficient, $a_{\beta\nu}$, and the Fierz interference term in Eq. (1). The latter parameter cannot be neglected when measuring the $\beta - \nu$ angular correlation coefficient, since it does not depend on any particular spin or momentum vector. Thus, it is more convenient to define a parameter $\tilde{a}$

$$\tilde{a} = \frac{a}{1 + b'} \tag{3}$$

where $b' \equiv \langle m/E_e \rangle b$ and $\langle\rangle$ stands for the weighted average over the observed part of the β spectrum [5]. However, within the SM, with only V and A interactions present, the Fierz interference term is equal to zero, such that $\tilde{a} \equiv a$. Then the $a_{\beta\nu}$ parameter can be expressed in terms of the Fermi and Gamow-Teller matrix elements:

$$\begin{aligned}a\xi =& \mid M_F \mid^2 \Big[- \mid C_S \mid^2 + \mid C_V \mid^2 - \mid C_S' \mid^2 + \mid C_V' \mid^2 \Big] + \\ & \frac{\mid M_{GT} \mid^2}{3} \Big[\mid C_T \mid^2 - \mid C_A \mid^2 + \mid C_T' \mid^2 - \mid C_A' \mid^2 \Big]\end{aligned} \tag{4}$$

Equation (4) highlights the sensitivity of the $a_{\beta\nu}$ parameter to possible scalar and tensor contributions, making it an interesting observable in the search for physics beyond the SM. This parameter depends quadratically on the coupling constants and thus, in order to get a high accuracy measurement and constrain its value, higher experimental precision is required.

If the decay rate can be measured as a function of the angle of emission between beta and neutrino, the correlation coefficients $a_{\beta\nu}$ and b can be constrained and a limit can be found on the relevant coupling constants. From Eq. (4) it becomes clear that the scalar and tensor interactions can be determined independently, i.e. by studying pure Fermi- and GT- transitions, given as result Eq. (5) and Eq. (6), respectively. Indeed, T and V make the largest $a_{\beta\nu}$ value when beta and neutrino are emitted in the same direction and hence result in a higher probability of small emission angles. On the other hand, S and A have the opposite effect. They reduce the $a_{\beta\nu}$ value and thus favour emission of beta and neutrino in opposite direction.

$$a_F \simeq 1 - \frac{\mid C_S \mid^2 + \mid C'_S \mid^2}{\mid C_V \mid^2} \tag{5}$$

$$a_{GT} \simeq -\frac{1}{3}\Big[1 - \frac{\mid C_T \mid^2 + \mid C'_T \mid^2}{\mid C_A \mid^2}\Big] \tag{6}$$

If one considers only SM contribution the previous parameters reduce to $a_F = 1$ and $a_{GT} = -\frac{1}{3}$. It is important to note here that the SM is actually quite good at describing the weak interaction. The current limits on the norm of the scalar coupling constant are $\mid C_S/C_V \mid = 0.0014(12)$ for left-handed currents and $\mid C_S/C_V \mid < 0.068$ (95% C.L.) for right-handed currents [5]. In order to improve this value, new measurements of the $a_{\beta\nu}$ parameter have to reach the 10^{-3} level of precision.

2.1. Motivation

A straightforward way to obtain experimental information about the $a_{\beta\nu}$ parameter would be to measure the correlation between the leptons emitted in the decay i.e. the β particle and the neutrino. Since a direct measurement of the neutrino is almost impossible, a common way to infer the $a_{\beta\nu}$ parameter is by performing high precision measurements of the recoiling daughter nuclei. However, this is also a difficult task, since the energy of the recoil is only a few hundred eV. In spite of that there is another insightful way to have information on $a_{\beta\nu}$; the beta-delayed proton emission. This requires using nuclei that are subject to β decay transitions and decay into the Isobaric Analogue State, *IAS*. An IAS is one that has a similar structure to the original decaying state allowing measurements between the β particle coming from the original decay and the emitted particle coming from the IAS state of the daughter nuclei. This technique was used by Schardt et al. [8] as a first observation of the $\beta - \nu$ recoil broadening for β delayed protons in ^{32}Ar and ^{33}Ar decay. This experiment was later improved by Adelberger et al. [7]. The most precise experiments performing measurements in the $a_{\beta\nu}$ parameter in search currents are presented in Table 1.

Table 1: Experimentally determined values for the $a_{\beta\nu}$ correlation coefficient for scalar contributions.

Isotope	$a_{\beta\nu}$	Technique	Reference
^{32}Ar	0.9989(52)	β^+ delayed p broadening	Adelberger et al. 1999 [7]
^{32}Ar	1.0050(52)*	^{32}Ar mass measurements	Blaum et al. 2003 [9]
^{38}K^m	0.9981(30)	β^+ delayed γ shift	Gorelov et al. 2005 [10]

* Using the exact measure of the ^{32}Ar mass, the previous $a_{\beta\nu}$ value obtained by Adelberger et al. was updated.

The values presented in Table 1 are in agreement with the expected value from the SM, $a_F = 1$, but the error bars still leave sufficient room for the possible existence of exotic currents.

The **WISArD** (**W**eak **I**nteraction **S**tudies with ^{32}**Ar**gon **D**ecay) experiment aims at a precise measurement of the $a_{\beta\nu}$ parameter using for the first time the kinematic energy shift of the β delayed protons emitted from the IAS of the ^{32}Ar ground state. Note that the measurements performed in the past by Schardt et. al. [8] and Adelberger et. al. [7], determined the broadening of these β delayed protons using the ^{32}Ar nuclide.

2.2. Experimental technique: The 0.1% challenge

The following section describes the proof-of-principle experiment carried out in the fall of 2018. We have decided to use radioactive argon isotope as our case of study due to the fact that the final state of the superallowed Fermi transition, the IAS in ^{32}Cl, is unbound to proton emission. A schematic diagram of this process is shown in Figure 1.

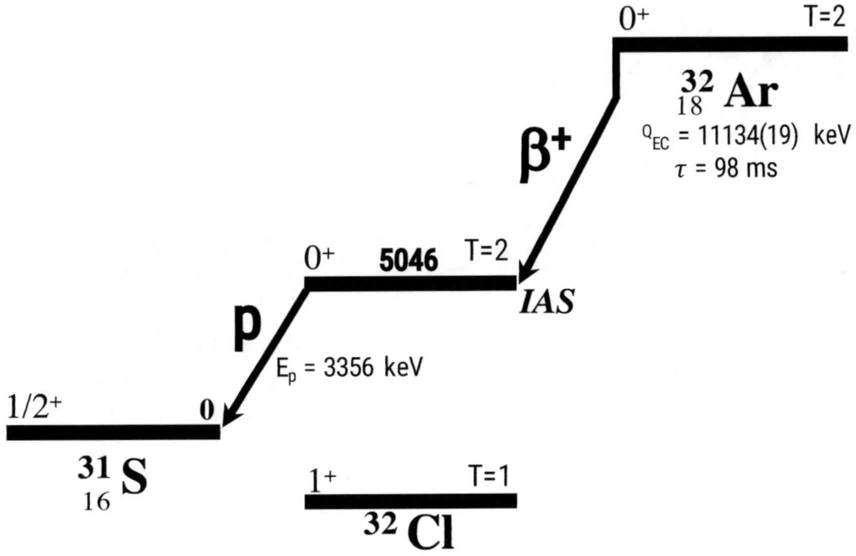

Figure 1: Schematic of the β-delayed proton decay of ^{32}Ar via the *IAS*.

Monte Carlo (*MC*) simulations were performed initially to test the feasibility of the experiment (see Figure 2) [12]. The simulated energy spectrum presents two cases for the protons emitted from the IAS in ^{32}Cl: (left) the dominant vector and (right) the exotic scalar contribution of the weak interaction. As one can see, there is a clear difference between the two currents. The vector current induces a kinematic shift on the proton energies (left side in Figure 2), whereas for a scalar current there is basically no difference with respect to the hemispheres in which the proton and positron are detected (right side in Figure 2).

This emphasizes that the energy resolution in the proton detectors will be crucial. If a proton resolution below 10 keV can be reached, a new limit for $a_{\beta\nu}$ of the order of 0.1% is reachable.

Due to the recoil the daughter nucleus gets from the emission of the positron and the neutrino, the proton is emitted from a moving source and is subject to a kinematic shift. If the proton is emitted in the direction in which the daughter nucleus moves it gets a boost in energy, (green and blue in Figure 2), whereas it is slowed down if emitted in opposite direction (red and pink, in Figure 2). Whatever is the case, measurements of the kinematic shift in the proton's energy is much easier to measure with high precision than a broadening. Keeping in mind that the correlation coefficient is sensitive to the energy and angle of the beta particle, we decided to measure positron-proton coincidences. Therefore, the detection setup requires a strong magnetic field to guide the positrons away from the proton detectors, reducing the background in the proton energy spectrum.

IOP Conf. Series: Journal of Physics: Conf. Series **1308** (2019) 012003 doi:10.1088/1742-6596/1308/1/012003

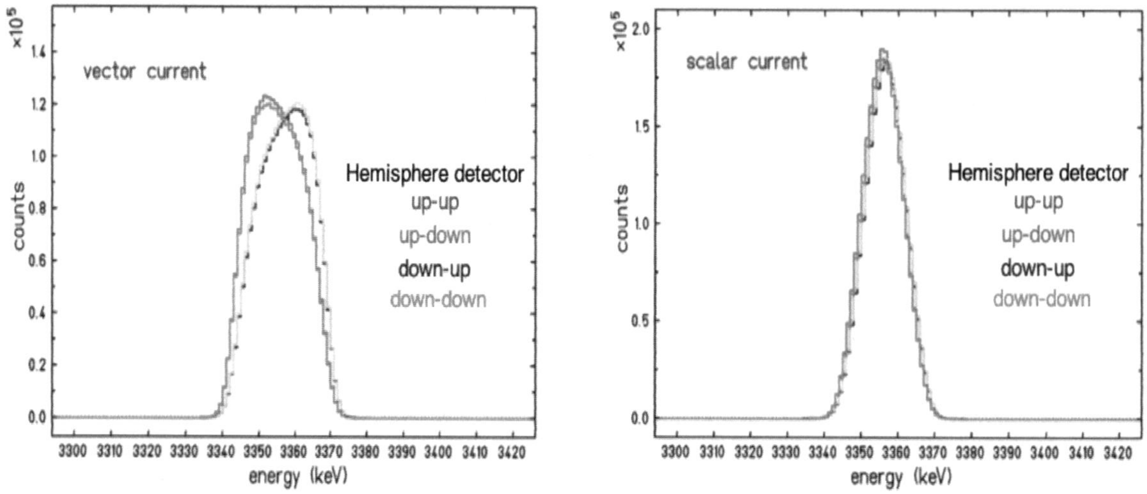

Figure 2: MC simulations showing the observed proton energy histogram from a pure vector and a pure scalar decay. The vector current induces a kinematic shift on the proton energies, whereas for a scalar current there is basically no difference with respect to the hemispheres in which the proton and the positron are detected.

3. WISArD setup

The new WISArD experiment reutilises the former WITCH superconducting magnet situated in the ISOLDE Hall at CERN. For this experiment a 1.4 GeV proton beam impinged on a thick CaO target. From the produced radioactive elements, the High Resolution Separator (HRS) selects the ^{32}Ar ions which are then send through REXtrap. From here, the continuous ^{32}Ar beam with an estimated intensity of the order of 2000 ions/s, got injected in the WISArD setup. After the beam traversed the horizontal beamline, the 30 keV ^{32}Ar beam is steered into the 7m high vertical beamline, where it is transported to the catcher foil situated at the geometrical center of a strong magnetic field (4T) generated by the superconducting magnet. The experimental setup is presented schematically in Figure 3. It shows how the ^{32}Ar beam is implanted continuously on a $6\mu m$ thick Mylar foil where the decay takes place. The protons emitted from the β delayed decay, slightly affected by the strong magnetic field, were detected on either side of the catcher foil using a mirrored configuration of an assembly of four silicon detectors. Positrons were guided by the magnetic field lines to a plastic scintillator mounted on the up plate downstream the mylar foil. Due to their bending radius, they cannot reach the proton detectors. In the first online proof-of-principle experiment, only one β detector was used. A second β detector located on the lower plate, upstream the mylar foil, is contemplated for future measurements.

The new detection system was expected to have a good energy resolution due to the low temperature inside the superconducting magnet. The signals were amplified by pre-amplifiers, which are located inside the superconducting magnet and developed to work under vacuum condition and strong magnetic fields, then digitized and recorded by the FASTER [13] data-acquisition system, which uses the ROOT Histogram Builder, *RHB*, for online visualization and data analysis. ROOT-trees contain single proton energy histograms and coincidence events between silicon and scintillator detectors. Data taken over a period of 20 hours using a 4T magnetic field are presented in the following section.

IOP Conf. Series: Journal of Physics: Conf. Series **1308** (2019) 012003 doi:10.1088/1742-6596/1308/1/012003

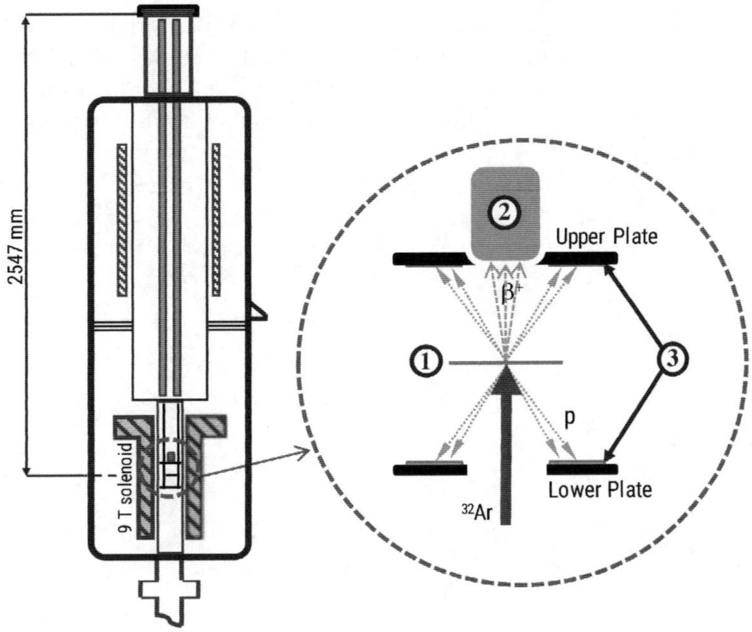

Figure 3: Schematic view of the WISArD magnet and detection setup. Left: the superconducting magnet of the former WITCH experiment in its cryostat. Notice the distance between the geometrical center of the detection setup (same as the geometrical center of the 9T solenoid), and the top flange, the latter being the only access to the system. Right: close-up of the detection setup in the center of the magnet. The 30 keV ^{32}Ar beam (purple arrow) is implanted in the mylar foil (1) where the decay takes place. Beta particles are guided by 4T magnetic field lines to the plastic scintillator (2) (Polystyrene) on the up plate. The Upper and Lower aluminium plates of 125 mm diameter and with a 20 mm hole in the center at both sides of the mylar foil are distance of 70 mm from the foil. Protons are detected with two disks of four round silicon detectors (3), 30 mm diameter each, fixed on these plates are installed. For the proof-of-principle experiment, only one β detector was installed.

4. Preliminary results

Preliminary results obtained from the campaign in November 2018 are presented.

When both leptons (i.e. positron and neutrino) emitted from the super-allowed Fermi β^+ decay, go in the same hemisphere, considering the mylar foil as a reference, they give to the recoiling nuclei a boost in energy. Following this approach, the kinematic energy shift will be more or less pronounced according to the direction of the emitted protons. Given that protons are emitted in any direction and are sensitive to the kinematic energy shift, we need to distinguish between the two directions of the recoiling daughter nuclei (i.e. the ^{32}Cl).

In order to evaluate the energy shift of the detected protons, we consider the data for one silicon detector in the upper and lower hemisphere. In the analysis hereafter we are considering only the protons emitted from the IAS at E_{lab} = 3356(2)keV. The single histogram was fitted with a Gaussian in the Fermi energy region. In Figure 4a, the Region Of Interest (ROI) in the Fermi energy peak is shown in with their close-up shown in 4b. The Fermi energy peak is expected at E_F = 3264.92(26) keV due to the energy loss in the mylar foil.

Figure 4c shows the protons that travel in the same the direction as the emitted β particles, (i.e. coincidence- histogram). Currently, the beta detector is located downstream the ^{32}Ar beam

IOP Conf. Series: Journal of Physics: Conf. Series **1308** (2019) 012003 doi:10.1088/1742-6596/1308/1/012003

(upper plate) allowing us to make coincidences between the β particles measured in the upper plate, (i) protons going to the upper plate, and (ii) protons going to the lower plate (opposite direction as the β emitted). The first case is presented in Figure 4c. In figure 4d, the close-up into the ROI Fermi energy peak is presented and fitted with a Gaussian yielding a mean energy value of E = 3259.95(36) keV.

The proton energy difference between the histograms in Figures 4b and 4d (i.e., single and coincidences in the upper hemisphere) is around 5 keV in the mean energy value. Expected reduction in the proton energy showed by the MC simulations (red in Figure 2).

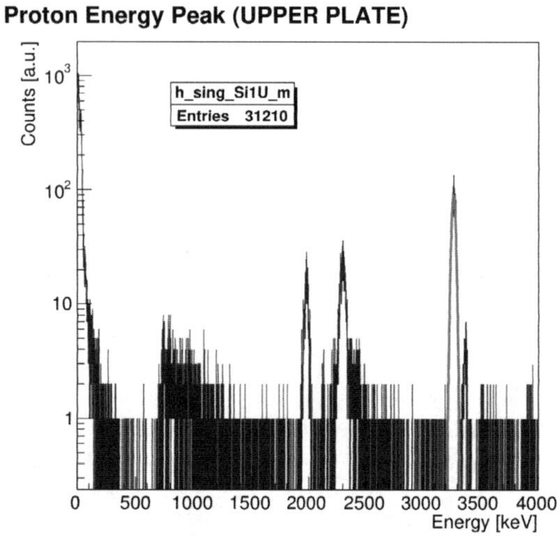

(4a) Proton energy spectrum fitted in the expected ROI (Fermi peak)

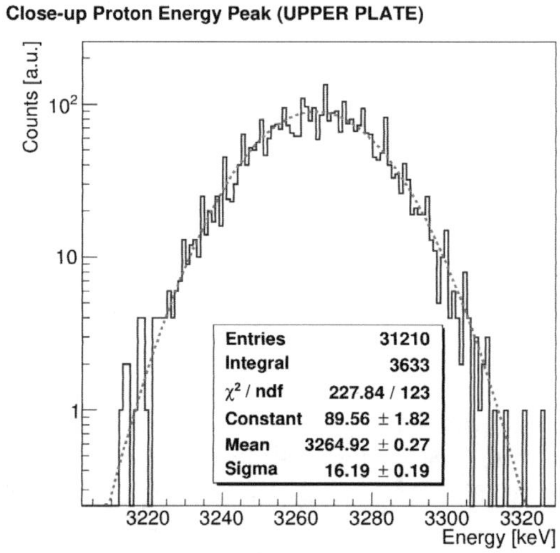

(4b) Close-up to the fit in (4a) yielding a mean value of E = 3264.92(26) keV

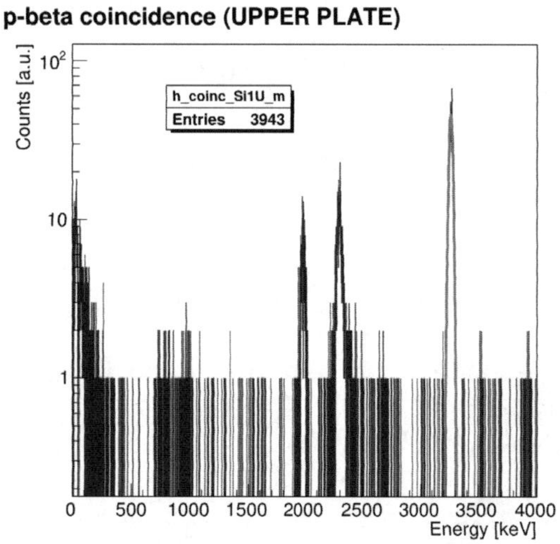

(4c) p-beta coincidence energy spectrum fitted in the expected ROI (Fermi peak)

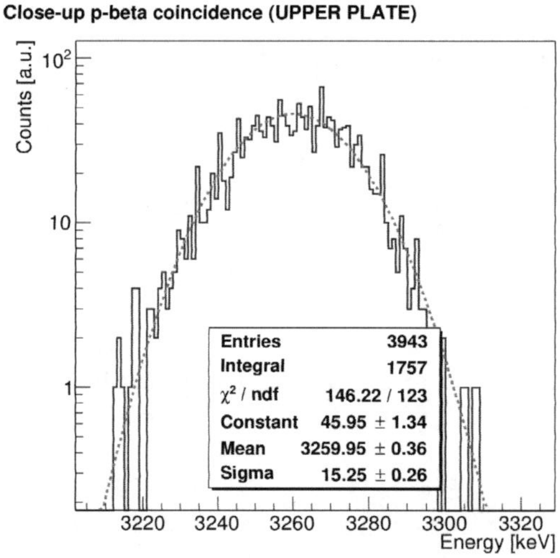

(4d) Close-up to the fit in (4c) yielding a mean value of E = 3259.95(36) keV.

IOP Conf. Series: Journal of Physics: Conf. Series **1308** (2019) 012003 doi:10.1088/1742-6596/1308/1/012003

A similar analysis is made taking the data from one of the four silicon detectors on the lower plate, shown in figure 5a, where the ROI in the Fermi energy peak and fitted (Gaussian) presented in 5b. Given a mean energy value of 3375.73(25) keV. In Figure 5c the second case is presented i.e., where the protons go into the opposite direction with respect to the beta particles. Figure **??** shows the close-up to the fit presented in the proton energy yielding a mean value of 3380.18(35) keV, corresponding to a difference in energy of about 4 keV with respect to the singles spectrum.

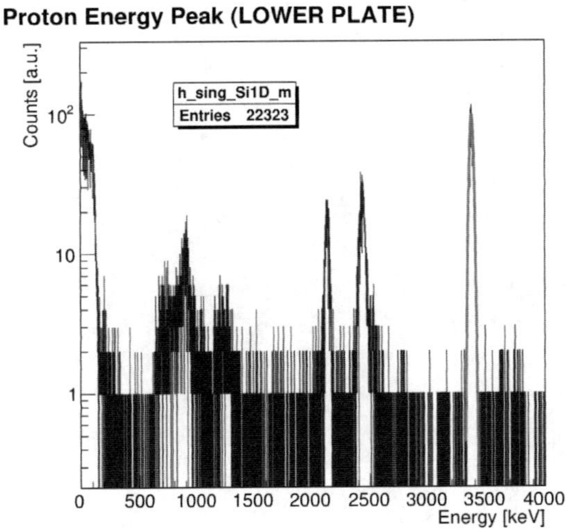

(5a) Proton energy spectrum fitted in the expected ROI (Fermi peak)

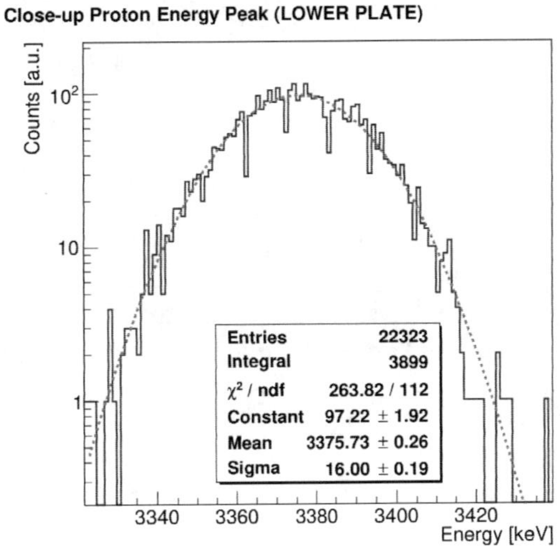

(5b) Close-up to the fit made in (5a) yielding a mean value of E = 3375.73(25) keV.

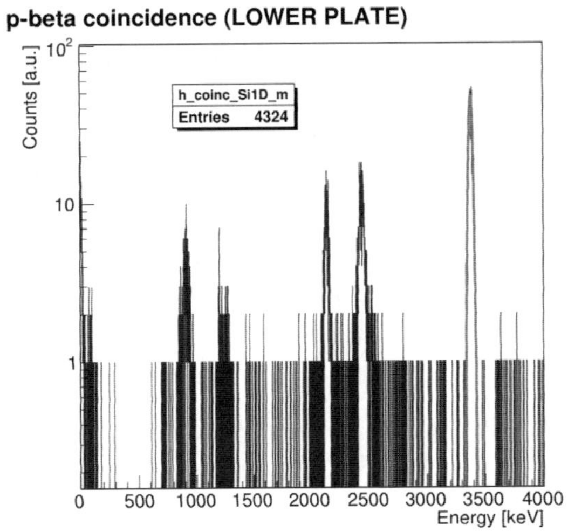

(5c) p-beta coincidence energy spectrum fitted in the expected ROI (Fermi peak)

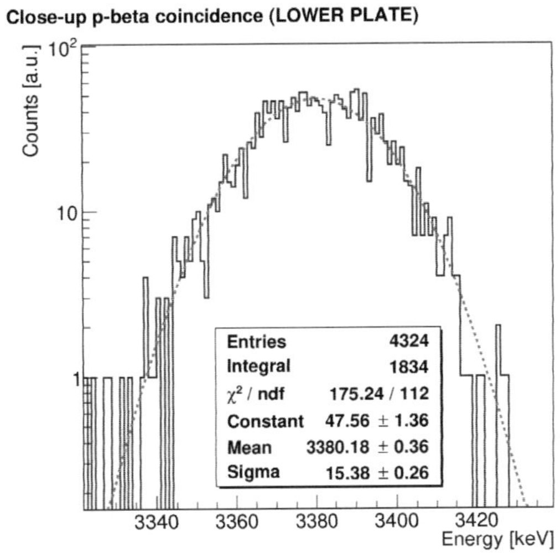

(5d) Close-up to the fit made in (5c) yielding a mean value of E = 3380.18(35) keV.

Table 2 summarizes the fitted mean values of a single ROOT file (2 hours of collected data) and the ROI used for each detector.

Table 2: Mean values for each spectrum in the Fermi peak.

	Silicon 1 UP		**Silicon 1 DW**	
case	ROI [keV]	Mean value	ROI [keV]	Mean value
Single	[3202 - 3316]	32**64**.92(26)	[3323 - 3438]	33**75**.73(25)
Coincidence	[3202 - 3316]	32**59**.95(36)	[3323 - 3438]	33**80**.18(35)

From the same analysis, it follows that the protons arriving in the Si1UP detector and so going in the same direction as the emitted leptons, also travelled through 5μm thick mylar foil. The implantation profile of ^{32}Ar ions at 30 keV energy in a mylar foil was computed with SRIM [14] showing that the sample will stop within the first 1μm. The mylar foil used in this first experiment was in the order of 6μm thick generating an energy loss around 100 keV. This difference is seen when comparing the mean energy values for the Si1UP and Si1DW. The second part of the Table 2 shows the mean energy values for the single- and coincidence- histograms for the Silicon 1 DW detector .

5. Outlook

Preliminary results show successfully the proof-of-principle of the experiment, since we were able to measure the proton kinematic energy shift, test the functionality of the new control system and optimize the beam injection into the superconducting magnet of the WISArD beamline. Previous MC simulations have shown that the energy resolution of the proton detectors is crucial (less than 10 keV FWHM) to achieve a 0.1% precision of $a_{\beta\nu}$ measurement. Our proton detectors gave an energy resolution in the order of 40 keV FWHM. Even though the kinematic energy shift is clearly seen, new developments and improvements for the energy resolution in the detection system will be done for new measurements, after the long shutdown at CERN in April 2021. Also, MC simulations are a vital component for understanding the systematic errors present in the experiment, i.e. energy loss due to mylar's thickness and backscattering in the plastic scintillator.

References

[1] Allen I S *et al* 1959 *Phys. Rev.* **116** 134
[2] González-Alonso M and O Naviliat Cuncic 2013 *Ann. Phys. (Berlin)* **525** 600
[3] Cirigliano V *et al* 2013 *Progr. Part. Nucl. Phys.* **71** 93
[4] Severijns N and Naviliat-Cuncic O 2013 *Phys. Scripta* **T152** 014018
[5] González-Alonso M, Naviliat-Cuncic O and Severijns N 2019 *Prog. Part. Nucl. Phys.* **104** 165
[6] Jackson J D, Treiman S B and Wyld H W 1957 *Phys. Rev.* **106(3)**, 517
[7] Adelberger E *et al* 1999 *Phys. Rev. Lett.* **83** 1299
[8] Schardt D and Riisager K 1993 *Z. Phys. A* **345** 265
[9] Blaum K *et al* 2003 *Phys. Rev. Lett.* **91** 26
[10] Gorelov A *et al* 2005 *Phys. Rev. Lett.* **94** 142501
[11] Vetter P A *et al* 2008 *Phys. Rev. C* **77**, 035502
[12] Blank B and Severijns N 2017 *J. Phys. G Nucl. Part. Phys.* **44** 074 002
[13] Fast Acquisition SysTem for nuclEar Research *FASTER* LPC-CAEN `http://faster.in2p3.fr`
[14] Ziegler J F *et al* 2010 *Nucl. Instr. and Meth. Phys. Res. B* **268** 11

Particle decay of astrophysically-important ^{19}Ne levels

D W Bardayan[1,2], K A Chipps[2,3,4], S Ahn[3,5], J C Blackmon[6], U Greife[4], K L Jones[3], A Kontos[5], R L Kozub[7], L Linhardt[6], B Manning[8], M Matos[2,3], P D O'Malley[1], S Ota[8], S D Pain[2], W A Peters[2,3], S T Pittman[2,3], A Sachs[3], K T Schmitt[2,3], M S Smith[2] and P Thompson[2,3]

[1] Dept. of Physics, University of Notre Dame, Notre Dame, IN 46556, USA
[2] Physics Division, Oak Ridge National Lab, Oak Ridge, TN 37831, USA
[3] Dept. of Physics and Astronomy, University of Tennessee, Knoxville, TN 37996 USA
[4] Dept. of Physics, Colorado School of Mines, Golden, CO 80401, USA
[5] National Superconducting Cyclotron Lab, Michigan State University, East Lansing, MI 48824, USA
[6] Dept. of Physics and Astronomy, Louisiana State University, Baton Rouge, LA 70803, USA
[7] Dept. of Physics, Tennessee Technological University, Cookeville, TN 38505, USA
[8] Dept. of Physics and Astronomy, Rutgers University, Piscataway, NJ 08854, USA

E-mail: danbardayan@nd.edu

Abstract. The ^{15}O$(\alpha,\gamma)^{19}$Ne reaction is an important trigger reaction leading to the rapid proton (rp) capture process in X-ray bursts. The primary uncertainty in determining its astrophysical rate is the uncertain α branching ratios of levels near $E_x = 4.1$ MeV in ^{19}Ne. These states have been populated in a study of the ^{20}Ne$(p,d)^{19}$Ne reaction, and α branching ratios are reported in this manuscript.

1. Introduction

The ^{15}O$(\alpha,\gamma)^{19}$Ne reaction triggers the rp-process in X-ray bursts and sets the stage for maximum light emission at the peak of the burst. The central role the ^{15}O$(\alpha,\gamma)^{19}$Ne reaction rate plays was recently demonstrated from its large effect on X-ray burst light curves [1]. In fact, if the ^{15}O$(\alpha,\gamma)^{19}$Ne reaction rate is near the lower limit of current estimates, models would even fail to predict X-ray burst explosions [2].

The temperatures at which the reaction rate becomes faster than the ^{15}O β decay is determined by the properties of ^{19}Ne levels near the α-emission threshold at 3.529 MeV. Of particular interest are levels at 4034 keV ($J^\pi = 3/2^+$), 4142 keV ($7/2^-$), and 4200 keV ($9/2^-$) [3]. While the levels energies are well known, it was only recently that the spins of the two higher-energy levels were clarified [4]. In addition, the lifetimes of these levels have been measured using Doppler-shift attenuation methods [5, 6]. The last remaining quantity that needs to be determined in order to calculate the ^{15}O$(\alpha,\gamma)^{19}$Ne rate are the α-branching ratios, $B_\alpha = \Gamma_\alpha/\Gamma$ since the α widths are expected to be much smaller than the γ widths. Multiple attempts have been made in the past to measure the α-branching ratios [3, 7, 8, 9] with one [3] reporting branching ratios of $B_\alpha = (2.9 \pm 2.1) \times 10^{-4}$ for the 4034-keV level and $B_\alpha = (1.2 \pm 0.5) \times 10^{-3}$ for

IOP Conf. Series: Journal of Physics: Conf. Series **1308** (2019) 012004 doi:10.1088/1742-6596/1308/1/012004

the combined 4142- and 4200-keV levels, which could not be resolved. All of the other previous measurements did not reach the sensitivity level required to measure the small α branching ratios near threshold. In this manuscript, a new analysis of ^{20}Ne$(p,d)^{19}$Ne data [10] is reported in order to study the α-branching ratios of astrophysically-important ^{19}Ne levels affecting estimates of the ^{15}O$(\alpha,\gamma)^{19}$Ne astrophysical reaction rate.

2. Experiment

A full description of the JENSA gas-jet target is given in Ref. [11]. Briefly, high pressure gas ($\sim$16000 Torr) is injected through a Laval nozzle to form a dense jet of gas approximately 4 mm wide. After passing through the target region, the gas is received, compressed, and recirculated back through the target with the aid of multiple stages of pumping and compression. The interaction point of the beam with the gas jet is surrounded by arrays of silicon detectors such as SIDAR [12], the Oak Ridge Rutgers University Barrel Array (ORRUBA) [13], and SuperORRUBA [14] The effective areal density of the jet was estimated by measuring the energy loss of α particles emitted from a ^{244}Cm source and was found to be $\sim 4\times10^{18}$ Ne atoms/cm^2. Precise knowledge of the target density was not necessary for this analysis as the primary goal was to measure the fraction of the time that a deuteron populating a given ^{19}Ne state was accompanied by a decay α particle.

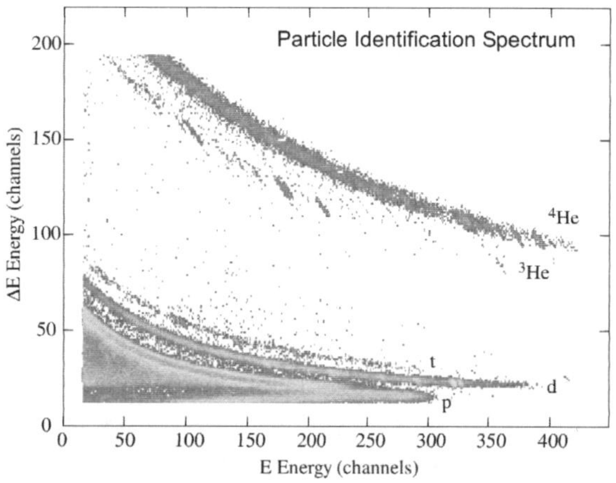

Figure 1. A particle identification spectrum from one of the SIDAR telescopes. The deuterons of interest are cleanly separated from other particle groups.

30-MeV proton beams were accelerated at the Holifield Radioactive Ion Beam Facility [15] and used to bombard ^{nat}Ne gas-jet targets. The beam was tuned and focused to a spot size of 2-3 mm at an optically-aligned retractable scintillating phosphor ensuring spatial overlap with the JENSA gas jet. Reaction ejectiles were detected and identified using elements of the SIDAR configured in telescope mode with 65-μm-thick ΔE detectors being backed by 1000-μm-thick E detectors and covering laboratory angles between $18^\circ - 53^\circ$. A particle-identification spectrum from the experiment is plotted in Fig. 1. The various observed particle groups ($p, d, t, ^3$He, and ^{4}He ions) were well separated using this energy-loss technique. The most intense group (10-20 kHz) arose from elastically-scattered protons, which were preferentially suppressed from the data acquisition by applying a hardware energy threshold to the ΔE detector signals.

The observed deuteron spectrum produced at 29° after bombardment with 3 nA proton beams for 15 hours is shown in Fig. 2. The spectrum shows excellent correspondence with known ^{19}Ne levels. A few deuteron peaks were observed from the ^{22}Ne$(p,d)^{21}$Ne reaction (labeled as ^{22}Ne in Fig. 2), but those were easily identified by their differing kinematic shifts. All other peaks correspond to known states in ^{19}Ne. The relatively smooth background between the peaks in

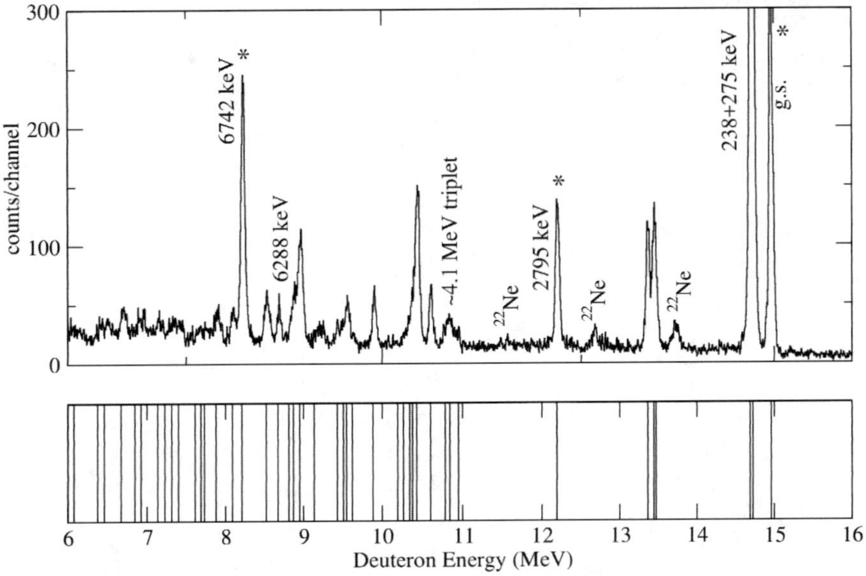

Figure 2. The top panel shows observed deuteron energy spectrum at 29° while the bottom panel shows the expected energies for the population of known levels in the ^{20}Ne$(p,d)^{19}$Ne reaction at the same angle and a bombarding energy of 30 MeV. Peaks labeled with stars were used for energy calibration.

the spectrum arises from effects such as incomplete charge collection in the silicon detectors and random coincidences in the deuteron gate arising primarily from pileup of elastically-scattered proton pulses. The deuteron energy calibration was performed using the strongly populated, isolated, and well-known levels at $E_x = 0$, 2795, and 6742 keV. Multiple states above the α and p thresholds were observed including the triplet of states near 4.1 MeV important for determining the ^{15}O$(\alpha,\gamma)^{19}$Ne reaction rate.

To search for α-decay events originating from the decay of excited ^{19}Ne levels, a spectrum was constructed of coincident charged particles that hit any other telescope besides the one in which the deuteron was detected. This condition was implemented to avoid confusing decay events with events that might mimic decay such as crosstalk in a single detector. The coincidence window was set in hardware to 4 μs for this experiment, and the expected α energies (> 500 keV) were above the hardware threshold of 300 keV. If charged particles were detected from the decay of excited states, there should be a clear correlation between the decay energies and the excited state populated (i.e., the lower deuteron energies corresponding to population of higher excited states result in higher charged-particle decay energies). The spectrum constructed from this coincidence requirement is plotted in Fig. 3. A clear band is visible showing the correlation expected for decay α events with the lowest energy α particles corresponding to deuterons populating states at the α threshold. The deuteron energy spectrum for events occurring within this decay band are plotted in Fig. 4.

In order to extract the α-branching ratios for populated ^{19}Ne levels, the random coincidence background first needed to be estimated. Random coincidences mostly arise as a result of the relatively high elastically-scattered proton rate ($\sim 10 - 20$ Hz) in the silicon detectors as evidenced in Fig. 3 by vertical streaks at fixed deuteron energies populating bound ^{19}Ne levels (e.g., near the deuteron energy channel of 370). This random coincidence background was most intense at the lowest α energies. To estimate the probability for a random coincidence, the

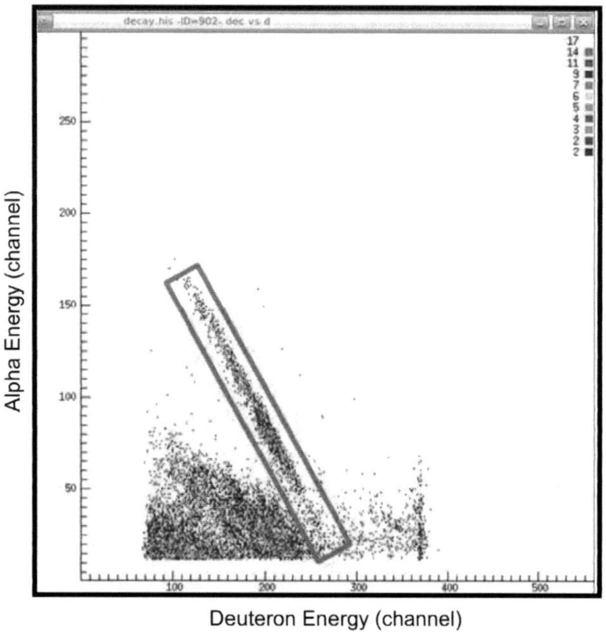

Figure 3. Decay energies plotted as a function of deuteron energies. Deuterons populating unbound ^{19}Ne levels show the expected correlation with decay α energies as indicated by the gate in this spectrum.

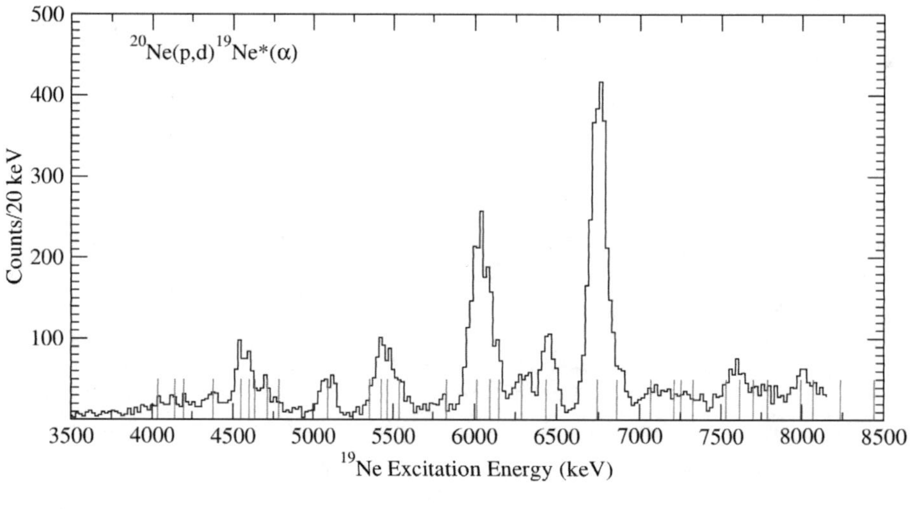

Figure 4. The deuteron energy spectrum in coincidence with decay α particles. As expected, decay α events are observed to start at the α threshold except for random coincidences producing a continuous background. The vertical lines indicate the previously-measured ^{19}Ne excitation energies. Small peaks are observed for the important 4.1-MeV triplet of states.

number of events were counted in coincidence with deuterons populating the bound ^{19}Ne levels at 0, 235, and 275 keV. The probability was observed to be as high as 0.8% between α energies of 400-800 keV and decreased to 0.005% between 3000-4000 keV.

The coincidence efficiency of the detector system was determined by considering the decay of ^{19}Ne levels known to decay with 100% α-branching ratios. The levels considered for this analysis were the 6742/6861-keV doublet with α widths 1000 times larger than any other decay widths [16]. After comparing the number of counts observed in this double peak in Figs. 2 and 4 and

IOP Conf. Series: Journal of Physics: Conf. Series **1308** (2019) 012004 doi:10.1088/1742-6596/1308/1/012004

subtracting the random coincidences ($\sim 0.01\%$), the coincidence efficiency was determined to be $(7.4 \pm 0.2)\%$.

Using these values for the random coincidence probabilities and the coincidence efficiency, the α branching ratios for other levels could be determined assuming isotropic distributions in the ^{19}Ne rest frame. The preliminary branching ratios measured from this work are shown in Table 1. The quoted uncertainties are statistical in nature and do not include the possible effects of anisotropic α distributions. At this stage, it has not been possible to extract α branching ratios for the ~4.1-MeV triplet of states as the random coincidence background is too large. It is hoped that through further analysis of the data, cleaner coincidence requirements can be implemented to further reduce this random coincidence background.

Table 1. Preliminary α branching ratios measured in this work. The excitation energies are taken from Ref. [10].

^{19}Ne Level	B_α (present)	B_α [9]	B_α [8]	B_α [7]
6.438(2) MeV	1.1 ± 0.1			
6.282(3) MeV	1.1 ± 0.2			
6.0-MeV triplet	1.00 ± 0.4			
5.539(9) MeV	0.88 ± 0.05			
5.090(6) MeV	0.61 ± 0.06	0.8 ± 0.1	0.90 ± 0.06	0.90 ± 0.09
4.371(3) MeV	$< 3 \times 10^{-2}$	0.016 ± 0.005	$< 3.9 \times 10^{-3}$	0.044 ± 0.032

3. Conclusions

Data from the ^{20}Ne$(p,d)^{19}$Ne reaction [10] have been re-analyzed to search for accompanying α decay. The α-decay branching ratios are important for constraining the astrophysical ^{15}O$(\alpha,\gamma)^{19}$Ne reaction rate. Branching ratios for higher-lying levels have been extracted but the sensitivity has not yet been reached in order to extract ratios for the most important 4.1-MeV triplet of ^{19}Ne levels. Further analysis of the data is in progress.

This work was supported in part by the U.S. Department of Energy under Contract Nos. DE-FG02-10ER41704 and DE-FG02-93ER40789 (CSM), DE-FG52-08NA28552 (Rutgers), DE-FG02-96ER40983 (UTK), DE-FG02-96ER40955 (TTU), DE-AC05-00OR22725 (ORNL), and the National Science Foundation under Contract Nos. PHY-1419765 (Notre Dame) and PHY-0822648 (JINA).

References

[1] J. L. Fisker, J. Gorres, M. Wiescher, Astrophys. J. **50**, 332 (2006).
[2] R. H. Cyburt, A. M. Amthor, A. Heger, E. Johnson, L. Keek, Z. Meisel, H. Schatz, K. Smith, Astrophys. J. **830**, 55 (2016).
[3] W. P. Tan, J. L. Fisker, J. Gorres, M. Couder, M. Wiescher, Phys. Rev. Lett. **98**, 242503 (2007).
[4] M. R. Hall *et al.*, Phys. Rev. C (in press).
[5] W. P. Tan *et al.*, Phys. Rev. C **72**, 041302(R) (2005).
[6] R. Kanungo *et al.*, Phys. Rev. C **74**, 045803 (2006).
[7] P. V. Magnus *et al.*, Nucl. Phys. A **506**, 332 (1990).
[8] B. Davids *et al.*, Phys. Rev. C **67**, 065808 (2003).
[9] K. E. Rehm *et al.*, Phys. Rev. C **67**, 065809 (2003).
[10] D. W. Bardayan *et al.*, Phys. Lett. B **751**, 311 (2015).
[11] K. A. Chipps *et al.*, Nucl. Instrum. Meth. Phys. Res. A **763**, 553 (2014).
[12] D. W. Bardayan *et al.*, Phys. Rev. Lett. **83** (1999) 45.
[13] S. D. Pain *et al.*, Nucl. Instr. and Meth. in Phys. Res. B **261** (2007) 1122.

[14] D. W. Bardayan *et al.*, Nucl. Instr. and Meth. in Phys. Res. A **711** (2013) 160.
[15] J. R. Beene *et al.*, J. Phys. G: Nucl. Part. Phys. **38** (2011) 024002.
[16] C. Akers *et al.*, Phys. Rev. C. **94**, 065803 (2016).

IOP Conf. Series: Journal of Physics: Conf. Series **1308** (2019) 012005 doi:10.1088/1742-6596/1308/1/012005

Odd-mass nuclei in the cluster shell model

R. Bijker and A.H. Santana-Valdés

Instituto de Ciencias Nucleares, Universidad Nacional Autónoma de México, A.P. 70-543, 04510 Ciudad de México, México

E-mail: bijker@nucleares.unam.mx

Abstract. In this contribution, we present the cluster shell model which is analogous to the Nilsson model, but for cluster potentials. Special attention is paid to the consequences of the discrete symmetries of three α-particles in an equilateral triangle configuration. This configuration is characterized by a special structure of the rotational bands which can be used as a fingerprint of the underlying geometric configuration. The cluster shell model is applied to the nucleus ^{13}C.

1. Introduction

Cluster degrees of freedom are very important for the description of light nuclei, in particular $k\alpha$ and $k\alpha + x$ nuclei, due to the large binding energy of the ^{4}He nucleus. Early work on α-cluster models goes back to the 1930's with studies by Wheeler [1], and Hafstad and Teller [2], followed by later work by Brink [3, 4] and Robson [5, 6]. Recently, there has been a lot of renewed interest in the structure of α-cluster nuclei, especially for the nucleus ^{12}C [7]. The measurement of new rotational excitations of the ground state [8, 9, 10] and of the Hoyle state [11, 12, 13, 14] has stimulated a large theoretical effort to understand the structure of ^{12}C (for a review see *e.g.* Refs. [7, 15, 16]).

In this contribution, we present a study of cluster states in ^{12}C and ^{13}C in the framework of the algebraic cluster model and the cluster shell model, respectively.

2. The algebraic cluster model

The algebraic cluster model (ACM) is an interacting boson model to describe the relative motion of cluster systems (see *e.g.* [17, 18]). The relevant degrees of freedom of a system of k-body clusters are given by the $k-1$ relative Jacobi coordinates and their conjugate momenta. The building blocks of the ACM consist of a vector boson for each relative coordinate and a scalar boson. Cluster states are then described in terms of a system of N interacting bosons with angular momentum and parity $L^P = 1^-$ (vector bosons) and $L^P = 0^+$ (scalar bosons). The $3(k-3)$ components of the vector bosons together with the scalar boson span a $(3k-2)$-dimensional space with group structure $U(3k-2)$. The many-body states are classified according to the totally symmetric irreducible representation $[N]$ of $U(3k-2)$. The ACM has a very rich symmetry structure. In addition to continuous symmetries like the angular momentum, in case of α-cluster nuclei the Hamiltonian has to be invariant under the permuation of the k identical α particles. Since one does not consider the excitations of the α particles themselves, the allowed cluster states have to be symmetric under the permutation group S_k.

Table 1. ACM for k-body clusters.

k	$U(3k-2)$	Symmetry	Geometry	Nucleus	Ref
2	$U(4)$	$Z_2 \sim S_2$	Dumbbell	^{8}Be	[19]
3	$U(7)$	$D_{3h} \supset D_3 \sim S_3$	Equilateral triangle	^{12}C	[20, 21]
4	$U(10)$	$T_d \sim S_4$	Regular tetrahedron	^{16}O	[22, 23]

The potential energy surface corresponding to the S_k invariant ACM Hamiltonian gives rise to several possible equilibrium shapes. The ones of special interest for applications to α-cluster nuclei, correspond to the geometrical configuration of a dumbbell for ^{8}Be ($k = 2$ with Z_2 symmetry), an equilateral triangle for ^{12}C ($k = 3$ with $D_{3h} \supset D_3$ symmetry) and a regular tetrahedron for ^{16}O ($k = 4$ with T_d symmetry), see Table 1. Even though these cases do not correspond to dynamical symmetries of the ACM Hamiltonian, one can still obtain approximate solutions for the rotation-vibration spectrum and electromagnetic properties [19, 21, 23].

In this contribution, we present the results for three-body clusters with a triangular configuration. The rotation-vibration spectrum is given by

$$E = \omega_1 \left(v_1 + \frac{1}{2} \right) + \omega_2 (v_2 + 1) + \kappa \, L(L+1) \; . \qquad (1)$$

The rotational structure of the ground-state band, $(v_1, v_2) = (0, 0)$, and the fundamental vibrations, $(1, 0)$ and $(0, 1)$, depends on the D_{3h} point group symmetry of the equilateral triangle configuration and is summarized in Fig. 1. The ground-state band is characterized by a rotational sequence involving both positive and negative parity states, $L^P = 0^+, 2^+, 3^-, 4^\pm, 5^-, \ldots$, all of which have been observed in ^{12}C. The so-called Hoyle band has the same structure, but so far only the positive parity states have been observed. The rotational bands in ^{12}C are shown in Fig. 2.

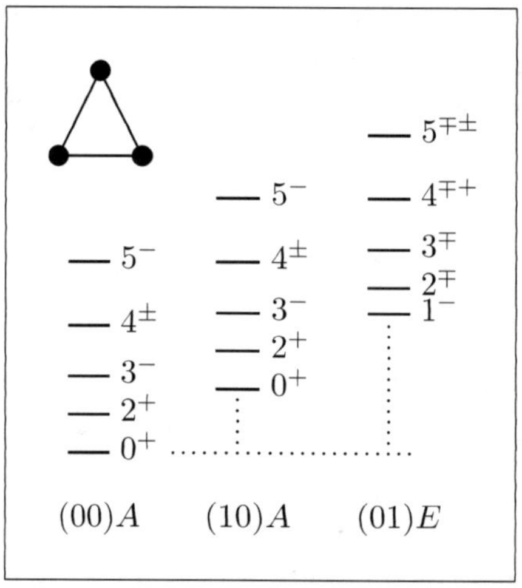

Figure 1. Schematic spectrum of a triangular configuration. The rotational bands are labeled by $(v_1 v_2)$ and t (bottom). All states are symmetric under $S_3 \sim D_3$.

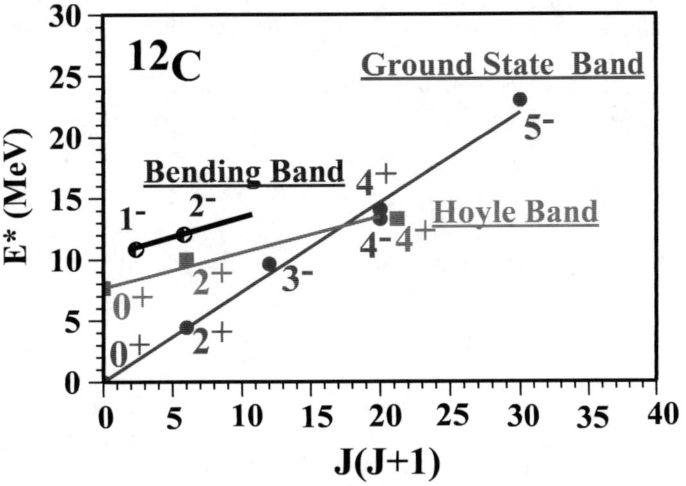

Figure 2. Rotational bands in ^{12}C [10].

The matter and charge distributions are usually taken as a Gaussian distribution

$$\rho(\vec{r}) = \left(\frac{\alpha}{\pi}\right)^{3/2} \sum_{i=1}^{3} \exp\left[-\alpha\left(\vec{r}-\vec{r_i}\right)^2\right] , \tag{2}$$

where the α-particles are located at a distance β from the center of mass with $\vec{r_i} = (\beta, \theta_i, \phi_i)$. The charge density is obtained by multiplying Eq. (2) by $Ze/3$ and the matter density by $Am/3$. Electromagnetic transition probabilities can be obtained from the transition form factors, which are the matrix elements of the Fourier transform of the charge distribution. For transitions along the ground-state band the transition form factors are given in terms of a product of a spherical Bessel function and an exponential factor arising from a Gaussian distribution of the electric charges [21]

$$F(0^+ \to L^P; q) = c_L\, j_L(q\beta)\, \mathrm{e}^{-q^2/4\alpha} . \tag{3}$$

The transition form factors depend on the parameters α and β. The value of α is determined from the radius of the α-particle to be $\alpha = 0.56$ fm^{-2} [24], and the value of β from the first minimum of the elastic form factor of ^{12}C to be $\beta = 1.74$ fm [21].

The transition probabilities $B(EL)$ along the ground-state band can be extracted from the form factors in the long wavelength limit

$$B(EL; 0^+ \to L^P) = \frac{(Ze\beta^L c_L)^2}{4\pi} , \tag{4}$$

with

$$c_L^2 = \frac{2L+1}{3}\left[1 + 2P_L(-\frac{1}{2})\right] \tag{5}$$

and the charge radius from the slope of the elastic form factor in the origin

$$\langle r^2 \rangle^{1/2} = \sqrt{\beta^2 + 3/2\alpha} . \tag{6}$$

Eq. (4) shows that all electric transitions are related to one another, and only depend on the value of β. For example, quadrupole, octupole and hexadecupole transitions are related by

$$\begin{aligned} \frac{B(E3;3_1^- \to 0_1^+)}{B(E2;2_1^+ \to 0_1^+)} &= \frac{5}{2}\beta^2 , \\ \frac{B(E4;4_1^+ \to 0_1^+)}{B(E2;2_1^+ \to 0_1^+)} &= \frac{9}{16}\beta^4 . \end{aligned} \tag{7}$$

The quadrupole moment can be calculated as [25]

$$Q_{2^+} = \frac{2}{7} Ze\beta^2 , \tag{8}$$

a positive value, as is to be expected for a planar configuration of three α-particles.

Table 2 shows a comparison between the experimental values of electromagnetic properties of ^{12}C and the ACM. According to Eqs. (4)-(8), the ACM results only depend on the value of α and β which have been determined from other observables, *i.e.* the radius of the α-particle and the first minimum of the elastic form factor. Therefore, the good agreement with the experimental values as shown in Table 2 provides evidence that the 2_1^+ and 3_1^- states indeed belong to the same ground-state rotational band of an equilateral triangle [21]. In particular, the large $B(E3)$ value cannot be obtained in shell-model calculations without the introduction of large effective charges.

Table 2. $B(EL)$ values, quadrupole moment and charge radius for ^{12}C. Experimental data are taken from [26, 27, 28], and the ACM values from [21].

	ACM	Exp	
$B(E2;2_1^+ \to 0_1^+)$	8.4	7.61 ± 0.42	e^2fm^4
$B(E3;3_1^- \to 0_1^+)$	73	104 ± 14	e^2fm^6
$B(E4;4_1^+ \to 0_1^+)$	44		e^2fm^8
$Q_{2_1^+}$	5.2	5.3 ± 4.4	efm^2
$\langle r^2 \rangle_{ch}^{1/2}$	2.389	2.468 ± 0.012	fm

3. The cluster shell model

For the description of odd-cluster nuclei, the cluster shell model (CSM) was developed recently [18, 19, 29]. The CSM combines cluster and single-particle degrees of freedom, and is very similar in spirit as the Nilsson model [30], but in the CSM the odd nucleon moves in the deformed field generated by the (collective) cluster degrees of freedom. The Hamiltonian is written as

$$H = T + V(\vec{r}) + V_{so}(\vec{r}) + V_C(\vec{r}) , \tag{9}$$

i.e. the sum of the kinetic energy, a central potential obtained by convoluting the density $\rho(\vec{r})$ of Eq. (2) with the interaction between the α-particle and the nucleon, a spin-orbit interaction and, for an odd proton, a Coulomb potential [29]. Fig. 3 shows the splitting of single-particles levels of a neutron moving in the deformed potential generated by a triangular configuration of three α-particles as a function of β. The solutions of the CSM Hamiltonian, χ_Ω with energy ϵ_Ω, are

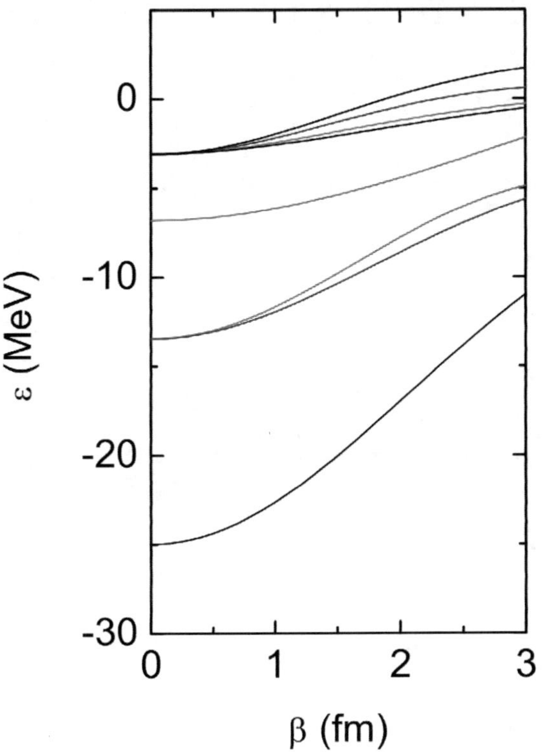

Figure 3. Single-particle energies in a cluster potential with $\mathcal{D}'_{3h}$ triangular symmetry. The single-particle levels are labeled by $E^{(+)}_{1/2}$ (black), $E^{(-)}_{1/2}$ (red) and $E_{3/2}$ (blue). For $\beta = 0$ the ordering of the single-particle orbits is $1s_{1/2}$, $1p_{3/2}$, $1p_{1/2}$ and (almost degenerate) $1d_{5/2}$, $2s_{1/2}$.

labeled by the irreducible representations of the double group D'_{3h}: $\Omega = E^{(-)}_{1/2}$, $E^{(+)}_{1/2}$ and $E_{3/2}$, each of which is doubly degenerate. The resolution of single-particle levels into representations of D'_{3h} is shown in Table 3.

In a study of ^{12}C the value of β was determined from the first minimum of the elastic form factor to be $\beta = 1.74$ fm [21]. Inspection of Fig. 3 shows that for this value of β, the first 6 neutrons occupy the intrinsic states with $\Omega = E^{(+)}_{1/2}$ (black), $E_{3/2}$ (blue) and $E^{(-)}_{1/2}$ (red), so that the last neutron in ^{13}C occupies the intrinsic state with $E^{(-)}_{1/2}$ (red), followed by $E^{(+)}_{1/2}$ (black).

The representations $E^{(+)}_{1/2}$, $E^{(-)}_{1/2}$ and $E_{3/2}$ can be further decomposed into values of K [31].

$$\begin{aligned}
\Omega = E^{(+)}_{1/2} \quad &: \quad K^P = \frac{1}{2}^+, \frac{5}{2}^-, \frac{7}{2}^-, \frac{11}{2}^+, \frac{13}{2}^+, \ldots \\
\Omega = E^{(-)}_{1/2} \quad &: \quad K^P = \frac{1}{2}^-, \frac{5}{2}^+, \frac{7}{2}^+, \frac{11}{2}^-, \frac{13}{2}^-, \ldots \\
\Omega = E_{3/2} \quad &: \quad K^P = \frac{3}{2}^\pm, \frac{9}{2}^\pm, \frac{15}{2}^\pm, \ldots
\end{aligned} \qquad (10)$$

with $n = 1, 2, 3, \ldots$, and $K > 0$. The angular momenta are given by $J = K, K+1, K+2, \ldots$.

IOP Conf. Series: Journal of Physics: Conf. Series **1308** (2019) 012005 doi:10.1088/1742-6596/1308/1/012005

Table 3. Resolution of single-particle levels into irreducible representations of D'_{3h}. Each E level is doubly degenerate.

	$E^{(+)}_{1/2}$	$E^{(-)}_{1/2}$	$E_{3/2}$
$s_{1/2}$	1	0	0
$p_{1/2}$	0	1	0
$p_{3/2}$	0	1	1
$d_{3/2}$	1	0	1
$d_{5/2}$	1	1	1
$f_{5/2}$	1	1	1
$f_{7/2}$	2	1	1

As a result, the rotational sequences for each one of the irreducible representations of D'_{3h} are given by (see also Table 3)

$$\begin{aligned}
\Omega = E^{(+)}_{1/2} \quad &: \quad J^P = \frac{1}{2}^+, \frac{3}{2}^+, \frac{5}{2}^\pm, \frac{7}{2}^+, \left(\frac{7}{2}^-\right)^2, \frac{9}{2}^+, \left(\frac{9}{2}^-\right)^2, \ldots \\
\Omega = E^{(-)}_{1/2} \quad &: \quad J^P = \frac{1}{2}^-, \frac{3}{2}^-, \frac{5}{2}^\pm, \left(\frac{7}{2}^+\right)^2, \frac{7}{2}^-, \left(\frac{9}{2}^+\right)^2, \frac{9}{2}^-, \ldots \\
\Omega = E_{3/2} \quad &: \quad J^P = \frac{3}{2}^\pm, \frac{5}{2}^\pm, \frac{7}{2}^\pm, \left(\frac{9}{2}^\pm\right)^2, \ldots
\end{aligned} \tag{11}$$

The angular momentum structure of each one of the representations of D'_{3h} is shown in Fig. 4.

The rotational energy spectra can be analyzed with

$$\begin{aligned}
E_{\rm rot}(\Omega, K, J) \quad = \quad & \varepsilon_\Omega + A_\Omega \Big[J(J+1) + b_\Omega K^2 \\
& + a_\Omega (-1)^{J+1/2} (J+1/2) \delta_{K,1/2} \Big] ,
\end{aligned} \tag{12}$$

where ε_Ω is the intrinsic energy, $A_\Omega = \hbar^2/2\mathcal{I}$ the inertial parameter, b_Ω a Coriolis term, and a_Ω the decoupling parameter. The latter term applies only to representations $\Omega = E^{(\pm)}_{1/2}$ and $K^P = 1/2^\pm$. Eq. (12) is the same energy formula as used by Nuhn in Ref. [32] in a description of the bandheads of ^{13}C in the context of a two-center shell model description of the system $^{13}\mathrm{C} + ^{16}\mathrm{O} \rightarrow ^{29}\mathrm{Si}$. Nuhn used an axially symmetric potential, in contrast to a cluster potential with triangular D_{3h} symmetry in the CSM. As a consequence, in the CSM the $K^P = 1/2^-$ and $5/2^+$ bands belong to the same configuration $\Omega = E^{(-)}_{1/2}$ (see Eq. (10) and Fig. 4), whereas in the axially symmetric case they represent separate rotational bands.

Fig. 5 shows the rotational bands of ^{13}C. The ground-state band has $K^P = 1/2^-$ and is assigned to the representation $\Omega = E^{(-)}_{1/2}$ of D'_{3h} (blue lines and filled circles) arising from the coupling of the ground-state band in ^{12}C to the intrinsic state with $E^{(-)}_{1/2}$. According to Eq. (10), this representation contains also $K^P = 5/2^+$ and $7/2^+$ bands, both of which appear to have been observed. In the shell model, positive parity states are expected to occur at much higher energies since they come from the s-d shell. The first excited rotational band has $K^P = 1/2^+$

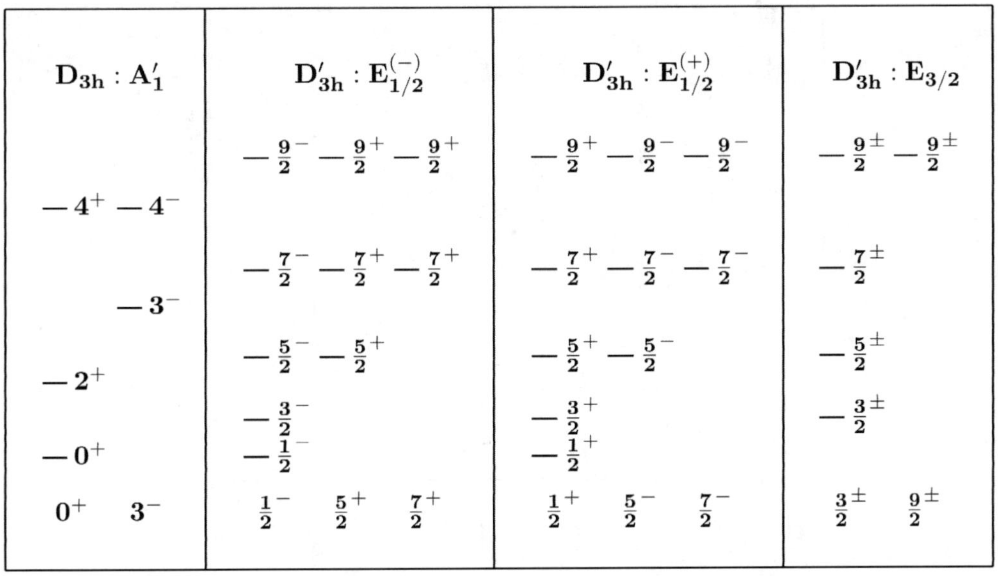

Figure 4. Structure of rotational bands for a triangular configuration of α particles in even-cluster nuclei (first panel) and odd-cluster nuclei with $E^{(-)}_{1/2}$, $E^{(+)}_{1/2}$ and $E_{3/2}$ symmetry (second, third and fourth panel). Each rotational band is labeled by the quantum numbers K^P.

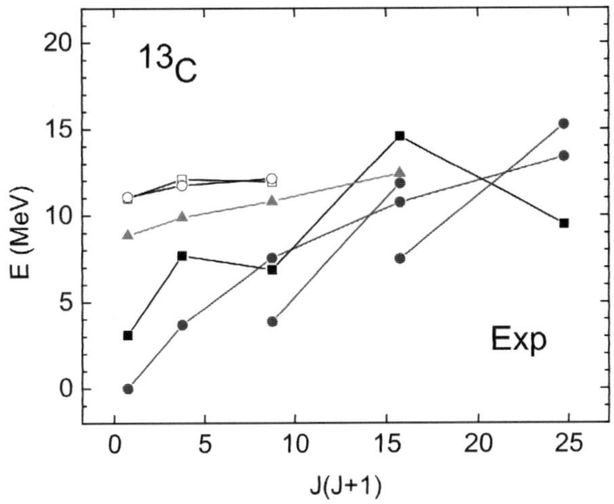

Figure 5. Rotational bands in ^{13}C [31].

which can be assigned to $\Omega = E^{(+)}_{1/2}$ (black line and filled squares) arising from the coupling of the ground-state band in ^{12}C to the excited intrinsic state with $E^{(+)}_{1/2}$. In contrast to the ground-state band, this excited band has a large decoupling parameter. In addition, Fig. 5 shows evidence for the occurrence of a rotational band at an energy slightly higher than that of the Hoyle state in ^{12}C which is interpreted as the coupling of the Hoyle band in ^{12}C to the ground-state intrinsic

IOP Conf. Series: Journal of Physics: Conf. Series **1308** (2019) 012005 doi:10.1088/1742-6596/1308/1/012005

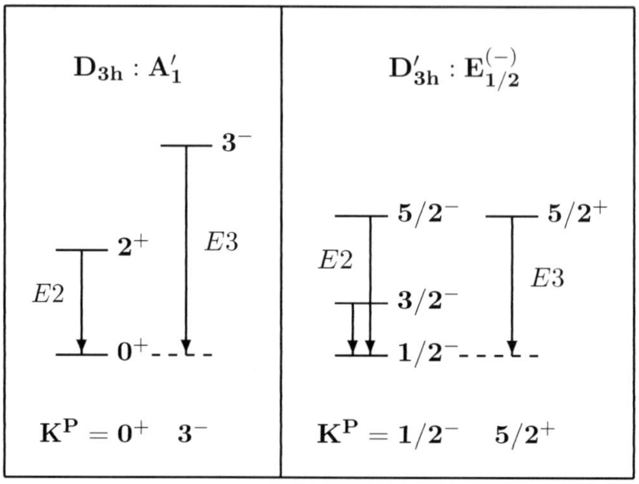

Figure 6. Electric transitions in ^{12}C (left) and ^{13}C (right).

state $E^{(-)}_{1/2}$ (red line and filled triangles).

Further evidence for the occurrence of D'_{3h} symmetry in ^{13}C is provided by electromagnetic transition rates. The $B(EL)$ values in ^{13}C are related to those in ^{12}C [31] (see also Fig. 6),

$$\begin{aligned} B(E2;3/2^-_1 \to 1/2^-_1) &= 0.99 \times B(E2;2^+_1 \to 0^+_1)\ , \\ B(E2;5/2^-_1 \to 1/2^-_1) &= 0.66 \times B(E2;2^+_1 \to 0^+_1)\ , \\ B(E3;5/2^+_1 \to 1/2^-_1) &= 0.57 \times B(E3;3^-_1 \to 0^+_1)\ . \end{aligned} \tag{13}$$

The numerical factors arise from combinations of Clebsch-Gordan coefficients. The results are shown in Table 4. The good agreement with the experimental $B(EL)$ values in ^{13}C shows that the states $3/2^-_1$, $5/2^-_1$ and $5/2^+_1$ belong to the same rotational band with $\Omega = E^{(-)}_{1/2}$. Just as in the case of ^{12}C there is a large octupole transition which is obtained in the CSM without the need to introduce effective charges.

Table 4. $B(EL)$ values in ^{13}C. Experimental data taken from [33], and the CSM values from [18, 31].

	CSM	Exp	
$B(E2;3/2^-_1 \to 1/2^-_1)$	8.3	6.4 ± 1.5	e^2fm^4
$B(E2;5/2^-_1 \to 1/2^-_1)$	5.5	5.6 ± 0.4	e^2fm^4
$B(E3;5/2^+_1 \to 1/2^-_1)$	42	100 ± 40	e^2fm^6

4. Summary and conclusions

In this contribution, we have presented a combined analysis of the rotation-vibration spectra and electromagnetic transition rates in ^{12}C and ^{13}C. A comparison with the theoretical predictions of the algebraic cluster model (^{12}C) and the cluster shell model (^{13}C) provides strong evidence for the occurrence of triangular symmetry in these nuclei. A characteristic feature of the triangular symmetry is the appearance of rotational bands consisting of both positive and negative parity states. The quadrupole and octupole transitions in ^{12}C and ^{13}C are strongly correlated and only depend on a single coefficient β whose value was determined independently from the first minimum in the elastic form factor of ^{12}C. The good agreement between theory and experiment supports the interpretation of the nucleus ^{13}C as a system of three α-particles in a triangular configuration plus an additional neutron moving in the deformed field generated by the cluster.

Acknowledgments
This work was supported in part by grant IN109017 from DGAPA-UNAM, Mexico.

References

[1] Wheeler J A 1937 *Phys. Rev.* **52** 1083
[2] Hafstad L R and Teller E 1938 *Phys. Rev.* **54** 681
[3] Brink D M 1965 *Int. School of Physics Enrico Fermi, Course XXXVI* 247
[4] Brink D M, Friedrich H, Weiguny A and Wong C W 1970 *Phys. Lett. B* **33** 143 the ACM and
[5] Robson D 1978 *Nucl. Phys. A* **308** 381
[6] Robson D 1982 *Prog. Part. Nucl. Phys.* **8** 257
[7] Freer M and Fynbo H O U 2014 *Prog. Part. Nucl. Phys.* **78** 1
[8] Freer M *et al.* 2007 *Phys. Rev. C* **76** 034320
[9] Kirsebom O S *et al.* 2010 *Phys. Rev. C* **81** 064313
[10] Marín-Lámbarri D J, Bijker R, Freer M, Gai M, Kokalova T, Parker D J and Wheldon C 2014 *Phys. Rev. Lett.* **113** 012502
[11] Itoh M *et al.* 2011 *Phys. Rev. C* **84** 054308
[12] Freer M *et al.* 2012 *Phys. Rev. C* **86** 034320
[13] Zimmerman W R *et al.* 2013 *Phys. Rev. Lett.* **110** 152502
[14] Freer M *et al.* 2011 *Phys. Rev. C* **83** 034314
[15] Schuck P, Funaki Y, Horiuchi H, Röpke G, Tohsaki A and Yamada T 2016 *Phys. Scr.* **91** 123001
[16] Freer M, Horiuchi H, Kanada-En'yo Y, Lee D and Meissner U G 2018 *Rev. Mod. Phys.* **90** 035004
[17] Bijker R 2016 *Phys. Scr.* **91** 073005
[18] Bijker R and Iachello F 2019 *Prog. Part. Nucl. Phys.* in press [arXiv:1903.04076]
[19] Della Rocca V and Iachello F 2017 *Nucl. Phys.* A **973** 1
[20] Bijker R and Iachello F 2000 *Phys. Rev.* C **61** 067305
[21] Bijker R and Iachello F 2002 *Ann. Phys. (N.Y.)* **298** 334
[22] Bijker R and Iachello F 2014 *Phys. Rev. Lett.* **112** 152501
[23] Bijker R and Iachello F 2017 *Nucl. Phys.* A **957** 154
[24] Sick I, McCarthy J S and Whitney R R 1976 *Phys. Lett. B* **64** 33
[25] Bijker R 2019 *AIP Conf Proc* in press [arXiv:1901.11057]
[26] Kelley J H, Purcell J E and Sheu C G 2017 *Nucl. Phys. A* **968** 71
[27] Kumar Raju M *et al.* 2018 *Phys. Lett. B* **777** 250
[28] Reuter W, Fricke G, Merle K and Miska H 1982 *Phys. Rev. C* **26** 806
[29] Della Rocca V, Bijker R and Iachello F 2017 *Nucl. Phys.* A **966** 158
[30] Nilsson S G 1955 *Kong Dan Vid Sel Mat Fys Med* **29** (16) 1
[31] Bijker R and Iachello F 2019 *Phys. Rev. Lett.* **122** 162501 [arXiv:1902.00451].
[32] Nuhn G, Scheid W and Park J Y 1987 *Phys. Rev. C* **35** 2146
[33] Ajzenberg-Selove F 1991 *Nucl. Phys. A* **523** 1

IOP Conf. Series: Journal of Physics: Conf. Series **1308** (2019) 012006 doi:10.1088/1742-6596/1308/1/012006

State of the art measurements with TexAT

J Bishop[1,2], G V Rogachev[1,2], E Aboud[1,2], S Ahn[1], M Assunção[7], M Barbui[1], A Bosh[1,2], V Guimarães[5], J Hooker[1,2], C Hunt[1,2], D Jayatissa[1,2], E Koshchiy[1], S Lukyanov[6], R O'Dwyer[1,2], Y Penionzhkevich[6], E Pollacco[4], C Pruitt[3], B T Roeder[1], A Saastamoinen[1], L Sobotka[3], E Uberseder[1], S Upadhyayula[1,2], J Zamora[5]

[1] Cyclotron Institute, Texas A&M University, College Station, TX 77843, USA
[2] Department of Physics & Astronomy, Texas A&M University, College Station, TX 77843, USA
[3] Department of Chemistry, Washington University, St. Louis, USA
[4] IRFU, CEA Saclay, Gif-sur-Yvette, France
[5] Universidade de São Paulo, Brazil
[6] Flerov Laboratory of Nuclear Reactions, JINR, Dubna, Russia
[7] Universidade Federal de São Paulo, Brazil

E-mail: jackbishop@tamu.edu

Abstract. In the past several years, there has been a large interest of Time Projection Chambers (TPCs) for use in experimental nuclear physics. This has continued in tandem with the requirement for high efficiency detectors with low intensity radioactive ion beams. TexAT is a Active Target TPC (AT-TPC) built at Texas A&M University utilizing MICRO MEsh GASeous (MicroMegas) pads and GET electronics developed specifically for TPCs. This design combines good TPC pixelation with a surrounding shell of Si/CsI telescopes to make an extremely versatile detector capable of a wide range of different experimental techniques with only minor modifications to the electronics setup. Two recent experiments performed at the Cyclotron Institute, Texas A&M University, are detailed here demonstrating versatility beyond the usual Thick Target Inverse Kinematics (TTIK) or transfer reactions that these TPCs are more typically used for. The first, a measurement of the $^{12}\mathrm{N} \to {}^{12}\mathrm{C}^{\star} \to 3\alpha$ decay demonstrates the capabilities of TexAT as a low-energy detector operating at low pressure (20 Torr) to measure β-delayed particle decay. The second, a direct measurement of the $^{8}\mathrm{B}+{}^{40}\mathrm{Ar}$ fusion cross section shows the advantages of operating in active target mode where the target also functions as the detector gas.

1. Introduction

Time projection chambers (TPCs) offer many advantages over conventional charged-particle spectroscopy setups (necessitating their use with low intensity radioactive ion beams) and have become increasingly common e.g. [1–7]. TPCs operate by filling a chamber with a drift gas. As charged particles lose energy in the gas and liberate electrons along their path, these electrons are collected by a uniform electric field and produce a signal. Measuring the time of arrival and position of this signal, the 3D path of the particles can be reconstructed. These drifted electrons however typically only have an energy of $\sim$ eV therefore require significant amplification before they can be detected. MICRO MEsh GASeous (MicroMegas) detectors [8, 9] achieve this by a

IOP Conf. Series: Journal of Physics: Conf. Series **1308** (2019) 012006 doi:10.1088/1742-6596/1308/1/012006

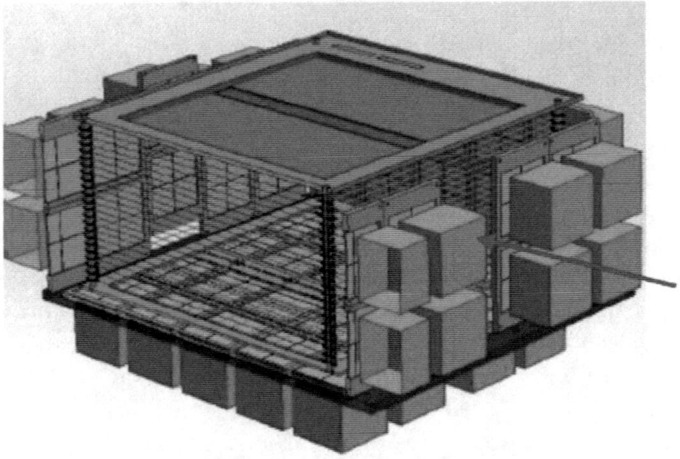

Figure 1. TexAT layout with the MicroMegas central pad (red) and side regions (green) surrounded by Si/CsI telescopes (cut-away view) following the future TexAT upgrade. The drift field is applied by the wires seen surrounding the active region.

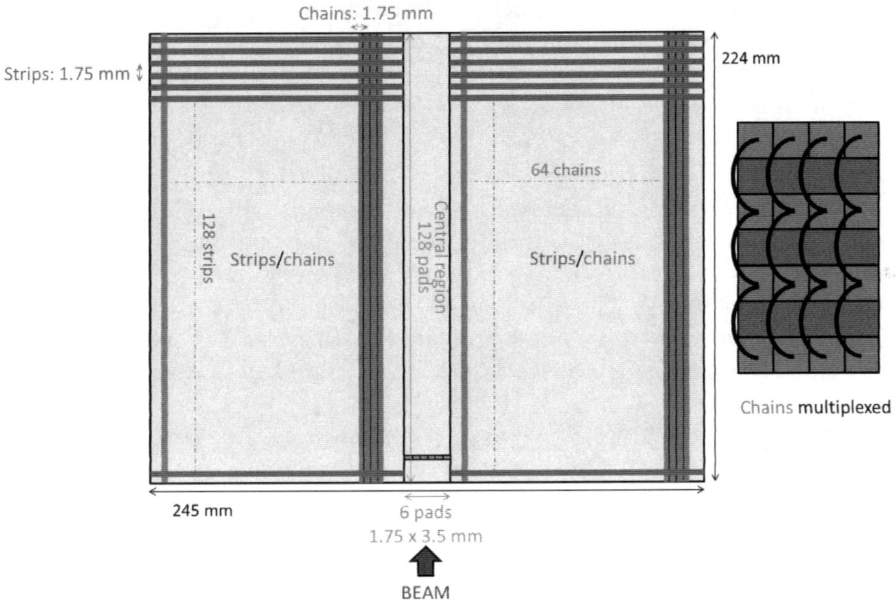

Figure 2. Layout of MicroMegas pads. There are two separate regions: the central region and the side regions with strips/chains. The central region contains 6 x 128 pads with a pitch of 1.75 mm in the beam direction and 3.5 mm in the perpendicular direction. These pads are read out individually. The side regions contains strips and chains which are multiplexed. The strips run perpendicular to the beam and run from the central region to the edge of the active area and constitute a continuous single strip. The chains are similarly multiplexed and run the length of this side region. These chains are spatially separated by the chains which run between them and are multiplexed as shown on the right by the black lines.

small amplification gap (128 μm) with a high voltage > 300 V. This large electric field generates a Townsend avalanche which can achieve a significant gain such that these drifted electrons can be detected. To maintain position sensitivity, the readout region is separated into pads with 1.75 mm pitch.

2. TexAT TPC

Fig. 1 shows TexAT, a TPC designed and built at the Cyclotron Institute, Texas A&M University [10–12]. The active region is 224 x 245 x 135 mm which is contained in an electric field provided by a field cage. At the top of the detector is a MicroMegas device which detects the drifted electrons created inside this active region. The pixelation provided by the MicroMegas pad is also shown in Fig. 2 where the increased pixelation in the centre region is apparent. This design also has a finer granularity in the beam direction. In the side regions, due to limitations in the number of readout channels, the pads require multiplexing for reading out. Horizontal (perpendicular to the beam direction) pads are read out together and are known as strips and the vertical pads (separated by the horizontal pads) are also multiplexed together and are known as chains. This multiplexing allows for a good granularity with a smaller number of readout channels required. Each readout channel is passed through into the GET electronics system [13] which has been specifically designed for TPCs. This system is shown in Fig. 3 where each channel can be seen to be fed into an AGET channel. This AGET chip digitises the signal at 25 MHz with 12 bit ADCs for 512 time buckets. Each channel also has an individual threshold which is fed into the MuTanT (MUltiplicity Trigger ANd Time) trigger unit.

3. Particle emission following β-decay

The flexibility of such a TPC as a probe of nuclear physics is vast. As a demonstration of this, an experiment was performed whereby states in ^{12}C above the α-decay threshold (7.37 MeV) were studied via the decay of ^{12}N. This β^+-decay has a half-life of 11.0 ms therefore must be investigated using the implantation technique rather than 'in-flight'. For this experiment, the main state of interest, the 0_2^+ resonance at 7.65 MeV known as the Hoyle state produces 3 α-particles which share 390 keV. This low energy therefore requires experimental care which is facilitated in this experiment by using a low pressure (20 Torr CO_2) fill gas in TexAT which gave the α-particles a long enough range that tracks can be measured. A decay-by-decay measurement

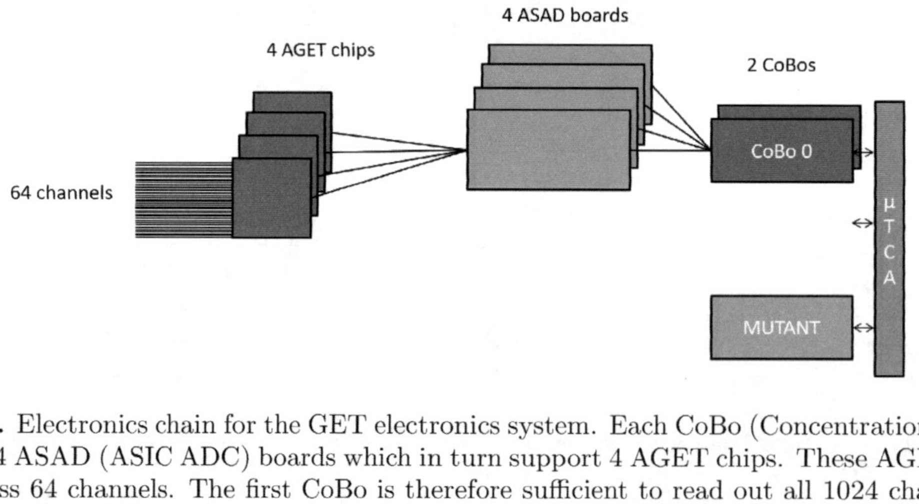

Figure 3. Electronics chain for the GET electronics system. Each CoBo (Concentration Board) supports 4 ASAD (ASIC ADC) boards which in turn support 4 AGET chips. These AGET chips can process 64 channels. The first CoBo is therefore sufficient to read out all 1024 channels of the MicroMegas allowing for 1024 channels from the second CoBo for Si/CsI detector channels.

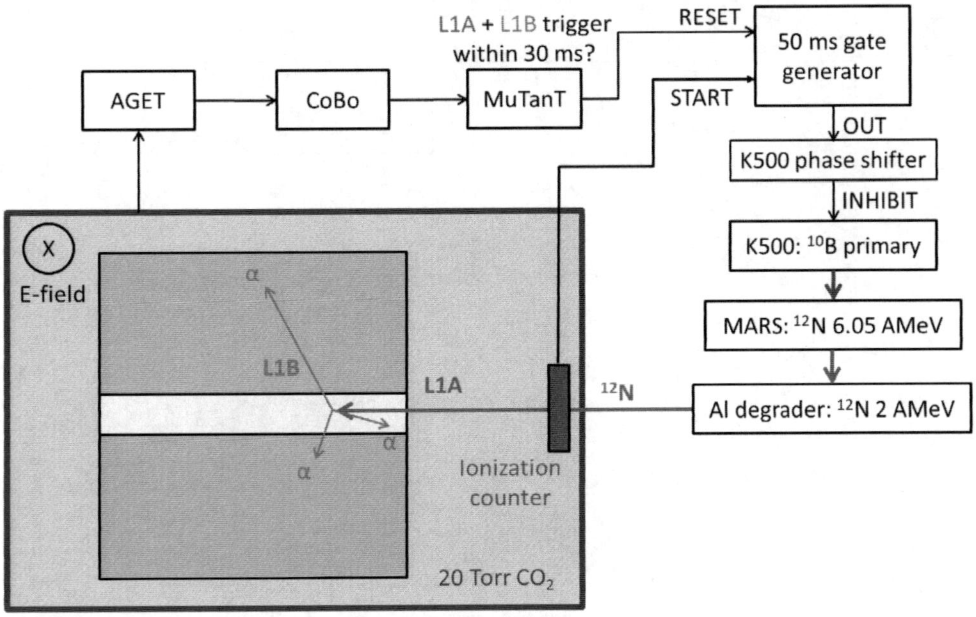

Figure 4. TexAT layout for the ^{12}N $\rightarrow 3\alpha$ experiment. The gate generator, triggered by the ion counter and reset by the successful measurement of a 3 α-particle decay feeds into the K500 cyclotron phase shifter. This allows for the beam to be stopped after an implantation has taken place.

can be achieved using the experimental setup shown in Fig. 4. This experiment was performed at the Cyclotron Institute, Texas A&M University, using the K500 cyclotron. The primary beam of ^{10}B then provided a ^{12}N secondary beam via the (^{3}He, n) reaction in the gas-cell on the MARS line. The secondary beam entered the detector at around 2 AMeV with a typical production intensity of 30 pps.

The measured data were split into two different frames using the 2p-mode in the GET electronics system [14] with two separate triggers. The first, the L1A trigger, corresponds to the track of the ^{12}N being implanted in the TPC and stopping. From this information, the decay position of the ^{12}N can be determined as well as verification via the energy deposited along its path that the track is a ^{12}N implant rather than a beam contaminant. The second trigger, the L1B, corresponds to the 3 α-particle decay of states in ^{12}C following the β^{+}-decay of ^{12}N . For both the L1A and the L1B, a pad multiplicity of > 1 is required for a duration of 10 timebuckets (a total of 400 ns) in the MicroMegas to generate this trigger. The TPC allows for 3D tracks of the three α-particles to be reconstructed. By measuring the total energy deposited, the excitation energy can be calculated by $E_x = E_{\text{measured}} + Q$. Additionally, selecting tracks where the range of the α-particles correspond to that expected from the decay of the Hoyle state (with the maximum α-particle energy being 190 keV), a clean selection on Hoyle decays can be achieved, as seen in Fig. 5 for a subset of the data. The counts excluded from this cut arise from higher lying states in ^{12}C which have higher energy α-particles which escape the MicroMegas and therefore do not deposit their full energy, instead creating a smooth distribution at higher energies.

IOP Conf. Series: Journal of Physics: Conf. Series **1308** (2019) 012006 doi:10.1088/1742-6596/1308/1/012006

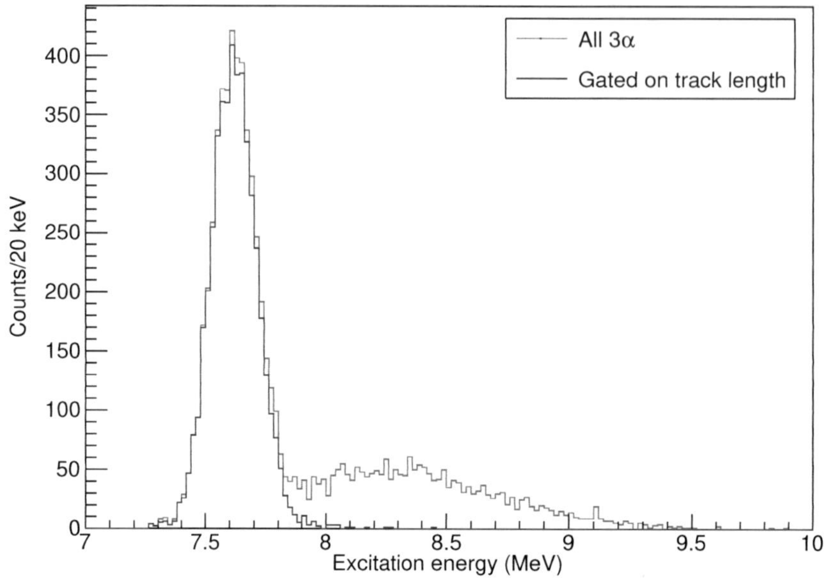

Figure 5. Excitation function for 3 α-particle events following the decay of ^{12}N calculated from the total energy deposited in the TPC. The gate corresponds to selecting events where the longest α-particle track is commensurate with a 190 keV energy corresponding to the decay of the Hoyle state into ^{8}Be(g.s) + α.

4. Direct fusion measurement

Traditional measurements of fusion cross sections across a wide energy range require measuring a particular exit channel and using models to then convert to the total cross section. By utilising a TPC as an active target, this fusion cross section can be measured directly across a wide range of energies. This technique relies on a measurement of the energy deposition as the impinging beam slows down in the target gas and then undergoes fusion. As the compound nucleus then splits into two (or more) fragments, the energy is then shared so as to conserve momentum. This spreading out of energy (in addition to the large increase in A and Z of the heavy recoil) means the energy deposition per unit length vastly increases. The identification of this vertex position then can be used to identify the centre of mass energy of the collision given the initial beam energy and the distance travelled in the gas. An accumulation of thousands of such events then gives a cross section measurement after correcting for the effective target thickness.

Of particular interest is the measurement of the ^{8}B + ^{40}Ar fusion cross section. As ^{8}B is a lightly-bound nucleus, believed to be a one-proton halo [15], the effect of this structure has an underlying effect on the fusion cross section as well as the ^{7}Be$(p,\gamma)^{8}$B reaction in main sequence stars. It has been found that the total fusion cross section is systematically lower for these weakly bound systems [16]. Careful investigation is therefore required to understand the role of coupling to the continuum.

The measurement of the ^{8}B + ^{40}Ar fusion cross section was therefore measured using TexAT with P_5 fill gas (95% Ar + 5% CH_4) at 150 Torr. To reduce the read-out rate of the TPC to avoid considerable dead-time for a high intensity beam such as ^{8}B, the MicroMegas were used in tandem with an ion counter placed at the entrance of the chamber. The signal of the ^{8}B passing through the ion counter provided a trigger for the acquisition while the last eighth of the

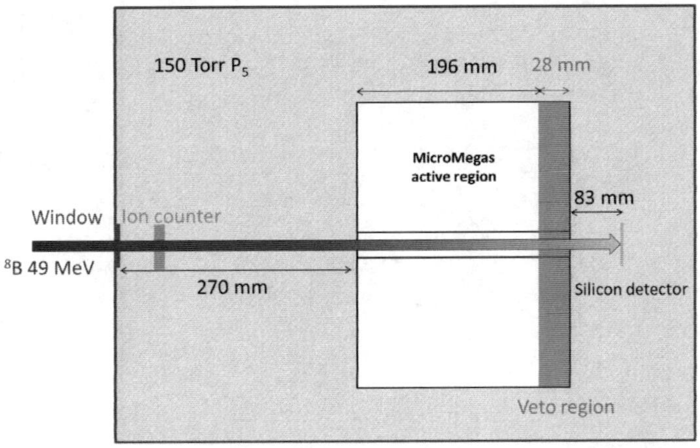

Figure 6. Experimental setup for direct fusion measurement. The TexAT chamber was filled with 150 Torr P5 gas. A 49 MeV ^{8}B enters the chamber and provides a trigger via an ion counter. The beam has sufficient energy to stop in a silicon detector placed at zero degrees. Upon traversing the last eight of the MicroMegas, a signal from this region is used to veto the ion counter signal. Therefore, only events which stop before the veto region are accepted. This removes non-interacting beam events and leaves breakup, fusion and elastic scattering events which are written to disk.

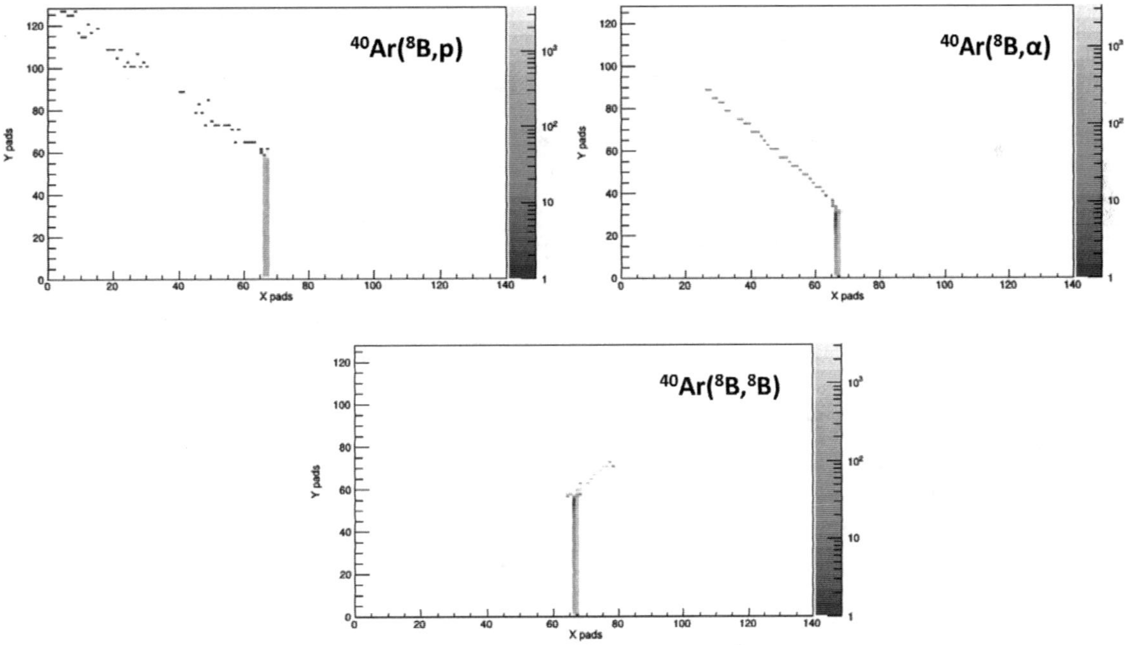

Figure 7. Top: GEANT4 simulated MicroMegas tracks. Top left: ^{40}Ar(^{8}B, p). Top right: ^{40}Ar(^{8}B, α). Bottom: ^{40}Ar(^{8}B, ^{8}B). In all these exit channels, the heavy recoil has a very short range and therefore deposits all of its energy at the interaction vertex. In the proton evaporation channel, the proton has enough energy to enter the veto region but is sufficiently high energy that the energy loss along the track is below the detection threshold therefore does not veto the event.

IOP Conf. Series: Journal of Physics: Conf. Series **1308** (2019) 012006 doi:10.1088/1742-6596/1308/1/012006

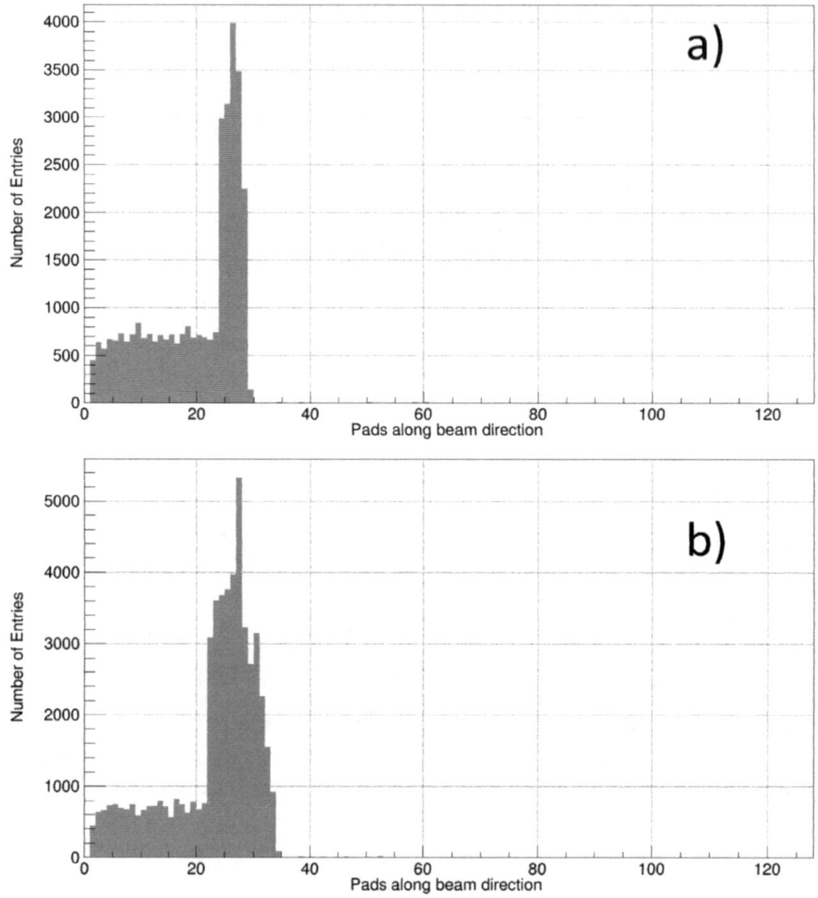

Figure 8. a) Energy loss as a function of distance traveled along the MicroMegas for the $^{40}Ar(^{8}B, ^{4}He)^{44}Sc$ reaction. b) Same plot but for $^{12}C(^{8}B, ^{4}He)^{16}F$ corresponding to interactions with the 5% methane in the P5 gas. While the peak signal is similar for the two processes, the width of the peak for the Ar interaction is narrower by virtue of the reduced range of the recoiling ^{44}Sc versus the ^{16}F.

MicroMegas allowed for a veto signal to be given to the readout. If the ^{8}B does not undergo an interaction with the gas (break-up/fusion/elastic scattering) then the beam has sufficient energy to traverse the MicroMegas and hit a silicon detector at zero degrees. This will therefore provide both a trigger (from the ion counter) and a veto (from the MicroMegas). If an interaction were to take place, then the outgoing particles will have lost sufficient energy that they will stop in the gas before reaching the veto region and the event is taken. Therefore, in this configuration, only events where the beam loses sufficient energy to not reach the end region of the detector are read out. This layout is summarized in Fig. 6.

This reaction was modelled in GEANT4 and can be seen in Fig. 7 for the proton - $^{40}Ar(^{8}B, p)$, α-particle - $^{40}Ar(^{8}B, \alpha)$ and compound-elastic channel - $^{40}Ar(^{8}B, ^{8}B)$. The proton channel is such that the proton can easily enter the veto region however due to the energy deposition per unit length (which is $\sim$ 40 times lower than the incoming beam), the proton track is below the detection threshold so does not veto the event.

IOP Conf. Series: Journal of Physics: Conf. Series **1308** (2019) 012006 doi:10.1088/1742-6596/1308/1/012006

Additionally, it is important to model the interaction of the beam with carbon in the methane which constitutes a background to the measurement. A separate run was performed with pure methane in order to allow for background subtraction. However on an event-by-event basis, the separation of interactions with argon or carbon was investigated. The energy loss along the beam direction is shown in Fig. 8 for these two scenarios with an interaction at the same position along the MicroMegas in the α-evaporation channel. The energy loss profile has a very similar peak size for the two interactions however the interaction with the argon has a much sharper peak due to the smaller range of the ^{44}Sc heavy recoil versus the recoiling ^{16}F for carbon fusion. By virtue of the mass disparity between the two, the fusion event with carbon also provides the heavy recoil with a larger total energy in comparison with the argon. As an example, for a 20 MeV ^{8}B undergoing fusion and exiting via the proton channel at $\theta_p = 55°$ in the lab, the argon fusion's heavy recoil (^{47}Ti) has an energy of 2.3 MeV in comparison with 6.2 MeV for the carbon fusion's heavy recoil (^{19}Ne). This can clearly be seen by the larger integral over the peak of Fig. 8 for the carbon interaction corresponding to a larger overall energy deposition.

5. Conclusion

As evidenced by these two studies, TexAT is able to facilitate a wide range of different techniques in addition to elastic scattering experiments most typically performed using this type of equipment. A combination of good central granularity of the MicroMegas coupled with sufficient gains allows for a good study of low energy-charged particles which may be used for a wide range of β-delayed charged particle experiments using the same 2p-mode implantation technique. Additionally, the GET electronics design allows for a versatile range of trigger conditions. This allows for high levels of trigger selection, without which, the TPC would become saturated with dead time. TexAT continues to be upgraded with a recently installed GEM layer [17] providing additional gain allowing for light particle tracks to be measured. Finally, the implementation of the higher coverage Si/CsI telescopes surrounding the MicroMegas region is currently under way, further increasing this device's capabilities.

References

[1] Hong B *et al* 2014 *Eur. Phys. J. A* **50** *49*
[2] Vorobyov A A *et al* 1988 *Nucl. Instrum. Methods Phys. Res. A* **270**, 419–30
[3] Roger T *et al* 2018 *Nucl. Instrum. Methods Phys. Res. A* **895**, 126–34
[4] Demonchy C E *et al* 2007 *Nucl. Instrum. Methods. Phys. Res. A* **583** 341–9
[5] Mizoi Y *et al* 1999 *Nucl. Instrum. Methods. Phys. Res. A* **431** 112–22
[6] Suzuki D *et al* 2012 *Nucl. Instrum. Methods. Phys. Res. A* **691** 39-54
[7] Beceiro-Novo S *et al* 2015 *Prog. Part. Nucl. Phys.* **84** 124–585
[8] Giomataris Y, *et al* 1996 *Nucl. Instrum. Methods Phys. Res. A* **376** 29-35
[9] Suzuki D, *et al* 2011 *Nucl. Instrum. Methods Phys. Res. A* **660** 64–8
[10] Koshchiy E *et al* Status of Texas Active Target (TexAT) detector
[11] Uberseder E *et al* Texas active target (TexAT) detector - part 2: Monte Carlo simulations
[12] Uberseder E *et al* Texas active target (TexAT) detector - part 3: Acquisition and analysis infrastructure
[13] Pollacco E C *et al* 2018 *Nucl. Instrum. Methods Phys. Res. A* **887**, 81-93
[14] Giovinazzo J *et al* 2007 *Phys. Rev. Lett.* **99**, 102501
[15] Minamisono T *et al* 1992 *Phys. Rev. Lett.* **69**, 2058
[16] Lei J and Moro A M 2019 *Phys. Rev. Lett.* **122**, 042503
[17] Kaminski J *et al* 2004 *Nucl. Instrum. Methods Phys. Res. A* **535** 201–5

IOP Conf. Series: Journal of Physics: Conf. Series **1308** (2019) 012007 doi:10.1088/1742-6596/1308/1/012007

Study of the ^{32}S(^{3}He,d)^{33}Cl one-proton transfer reaction with a new generation hodoscope

M Cinausero[1], D Dell'Aquila[2], I Lombardo[2], M Vigilante[3,4], S Barlini[5], R Bolzonella[10,1], M Bruno[6,7], A Buccola[8,5], S Cartuan[1], G Casini[5], M Cicerchia[10,1], M D'Andrea[2], M Degerlier[12], D Fabris[9], L Gasques[11], F Gramegna[1], A Lepine-Szily[11], G Mantovani[10,1], T Marchi[1], S Piantelli[5], V Rigato[1], L Scomparin[10,1] and S Valdrè[5]

[1] INFN-Laboratori Nazionali di Legnaro, Legnaro, Italy
[2] INFN-Sezione di Catania, Catania, Italy
[3] Università degli Studi di Napoli "Federico II", Napoli, Italy
[4] INFN-Sezione di Napoli, Napoli, Italy
[5] INFN-Sezione di Firenze, Firenze, Italy
[6] Università degli Studi di Bologna, Bologna, Italy
[7] INFN-Sezione di Bologna, Bologna, Italy
[8] Università degli Studi di Firenze, Firenze, Italy
[9] INFN-Sezione di Padova, Padova, Italy
[10] Università degli Studi di Padova, Padova, Italy
[11] Instituto de Fisica da Universidade de São Paulo, 05508-090 São Paulo, Brasil
[12] Science and Art Faculty, Physics Department, Nevsehir Haci Bektas Veli Univ., Nevsehir, Turkey

E-mail: dellaquila.daniele@gmail.com

Abstract. The ^{32}S(^{3}He,d)^{33}Cl one-proton transfer reaction is a powerful tool to investigate the spectroscopy of low-lying states in the proton-rich ^{33}Cl nucleus. However, the extraction of firm differential cross-section data at various angles, against which benchmarking theoretical models to correctly constrain the spectroscopy of ^{33}Cl, is made challenging by the presence of competitive reaction products contaminating the detected energy spectra. We have recently measured the ^{32}S(^{3}He,d)^{33}Cl reaction at 9.8 MeV incident energy by using a new generation hodoscope of silicon detectors, capable to detect and identify emitted deuterons down to energies of the order of 2 MeV. The high angular segmentation of our hodoscope allowed to unambiguously disentangle the contribution of one-proton transfer reactions in the ground state of ^{33}Cl and in its 0.810 MeV, 2.352 MeV, 2.685 MeV, 2.846 MeV excited states from contaminant deuteron-emitting reactions. These data will be crucial to help to constrain J^{π} and spectroscopic factor C^2S_p values of low-lying ^{33}Cl states, still ambiguous in the literature. The present status of the analysis is discussed in the paper.

1. Introduction

Nuclear reactions at low and intermediate energies allow to probe many properties of nuclear systems and their governing forces such as the structure of nuclei [1, 2, 3, 4] and clustering phenomena [5, 6, 7], their influence in Astrophysics [8], the dynamics of nucleus-nucleus collisions [9, 10, 11], and their interplay [12]. While the latter requires often the use of 4π detectors [13, 14, 15] because of the complexity of the final state of the collisions, nuclear spectroscopy studies are typically more selective, enabling the use of more compact devices [16]. In this framework, transfer reactions are a class of particularly selective processes that allow to explore specific degrees of freedom in nuclear systems such as single-particle configurations. In the literature, several transfer reaction studies have been performed by using spectrometers like MAGNEX at INFN-LNS [17], VAMOS at GANIL [18] and K600 at iThemba LABS [19]. Recently, new high-resolution and modular detectors for nuclear physics have been developed (see, for example FARCOS at INFN-LNS [20] or HiRA10 at MSU-NSCL [21]) to be used for nuclear reactions or as ancillary systems for large acceptance devices [22, 23] or magnetic spectrometers [18, 24]. Such detectors have the advantage of high angular segmentation, high energy resolution and a high degree of modularity. In this paper, we exploit the new modular device OSCAR [25] to investigate the one-proton transfer reaction ^{32}S(^{3}He,d)^{33}Cl at 9.8 MeV incident energy. This reaction is particularly suited to probe the spectroscopy (i.e. energy, J^{π} and C^2S_p spectroscopic factors) of low-lying states of the residual proton-rich ^{33}Cl nucleus by comparing measured differential cross section angular distributions with the predictions of theoretical calculations [26, 27]. However, while such theoretical predictions reached a level of good consistency, from the experimental point of view several ambiguities are still affecting the spectroscopy of ^{33}Cl low lying states, concerning especially J^{π} assignments and C^2S_p spectroscopic factors [28]. Such ambiguities arise especially from the presence of contaminants in the observed reaction products spectra that make it difficult to extract the differential cross section angular distributions in wide angular regions [29]. To improve previous studies of the ^{32}S(^{3}He,d)^{33}Cl reactions [30, 31], an essential ingredient is therefore to reduce possible sources of contaminants produced by undesired elements in the target and, additionally, to improve energy and angular resolution to more effectively disentangle the background from the signals to measure. Finally, a reliable measurement of the absolute cross-section scale is mandatory for a reasonable estimate of the C^2S_p factors.

2. Experimental setup

The experiment has been performed at the INFN-Laboratori Nazionali di Legnaro (LNL), Italy. A ^{3}He^{++} beam was delivered by the CN accelerator at an energy of 9.8 MeV on a 53 $\mu g/cm^2$ ^{nat}Zn^{32}S target, 99% enriched in ^{32}S. The ^{nat}Zn^{32}S powder is evaporated by electron beam technique on a thin substrate of carbon ($\approx$ 15 $\mu g/cm^2$) to reduce contaminant reactions. The target was installed in the center of a circular scattering chamber, shown in Figure 1. A rotating plate hosted two silicon telescopes, each constituted by an ORTEC surface barrier silicon followed by a high resolution Hamamatsu PIN diode, and an OSCAR telescope [25], placed on the opposite side of the beam line as shown in Figure 1. Both detectors have 2 detection stages to enable the particle identification via the ΔE-E technique. The OSCAR telescope was composed by a thin ($\approx$ 20 μm) 16-strips Single-Side Silicon Strip Detector (SSSSD) manufactured by Micron Semiconductors used as the first detection stage, and 16 individual silicon PIN diode pads (300 μm thick) manufactured by Hamamatsu used as the second detection stage. Each strip of the first detection stage is a 3.125 mm wide vertical silicon strip, while the silicon pads used as the second detection stage have an active area of 1 cm^2 and a ceramic frame of 1.4 mm and 3.2 mm on the bottom side. Each pad is installed on an electronic board containing plug-in pre-amplifiers, as discussed in [25] more in detail, while for the first detection stage we used a dedicated 16 channels pre-amplifier which lies below the rotating plate in Figure 1.

To determine the angular position of each detector with respect to the beam line we performed a dedicated measurement using an optical alignment telescope. Each of the measured positions is determined with an accuracy better than 0.2°. The optical alignment procedure allowed us also to benchmark the geometry information of the SSSSD detector claimed by the manufacturer. In the OSCAR hodoscope, the correlations between the first and second detection stages can be used to constrain with high precision the impact point on the surface of the detector and

Figure 1. A photography of the experimental setup used for the experiment at INFN-LNL. The OSCAR hodoscope is installed at about 12 cm from the target in a circular vacuum chamber available in the experimental hall of the 5.5 MV CN accelerator. Other two individual telescopes were additionally installed. The rotating plate allows to easily change the laboratory angles of each detector.

therefore the emission angle of the particle emitted from the target. We define 64 individual pseudo-telescopes, each identified by the correspondence of a particular strip of the SSSSD with a pad in the second detection stage. This method allows to have 64 detection telescopes using only 32 individual electronics channels. The detection angle associated to each pseudo-telescope is defined as the geometrical center of the corresponding strip-pad correspondence. These are calculated with a Monte Carlo tool, starting from the absolute position of the OSCAR hodoscope measured with the optical telescope. In the procedure we generate random directions of incidence in the OSCAR telescope and then calculate the azimuthal and polar angle center of each pseudo-telescope from the corresponding distributions. Effective solid angles of each pseudo-telescope were benchmarked by means of a dedicated Monte Carlo tool as described in Ref. [25]. During the experiment, we used two different position of OSCAR in order to cover a larger region of detection angles, improved with respect to the state-of-the-art literature [30, 31]. In the first of the two geometry positions, OSCAR covered the laboratory angles from about $\theta = 20°$ to about $\theta = 40°$, while in the second the laboratory angles from $\theta = 40°$ to $\theta = 60°$ are covered. The overall coverage of the cumulative data is shown in Figure 3 in a deuteron energy vs θ plot.

3. Status of the data analysis

Particles in the OSCAR hodoscope are reconstructed by correlating strips in the first detection stage with pads in the second detection stage, according to the algorithm described in Ref. [25]. The energy deposited in the detector is calibrated independently for the first and the second stage by using the elastic scattering $^3\text{He} +^{197}\text{Au}$ at 13 different incident energies, collected at the end of the experimental run. To account for small electronic drift effects affecting the

IOP Conf. Series: Journal of Physics: Conf. Series **1308** (2019) 012007 doi:10.1088/1742-6596/1308/1/012007

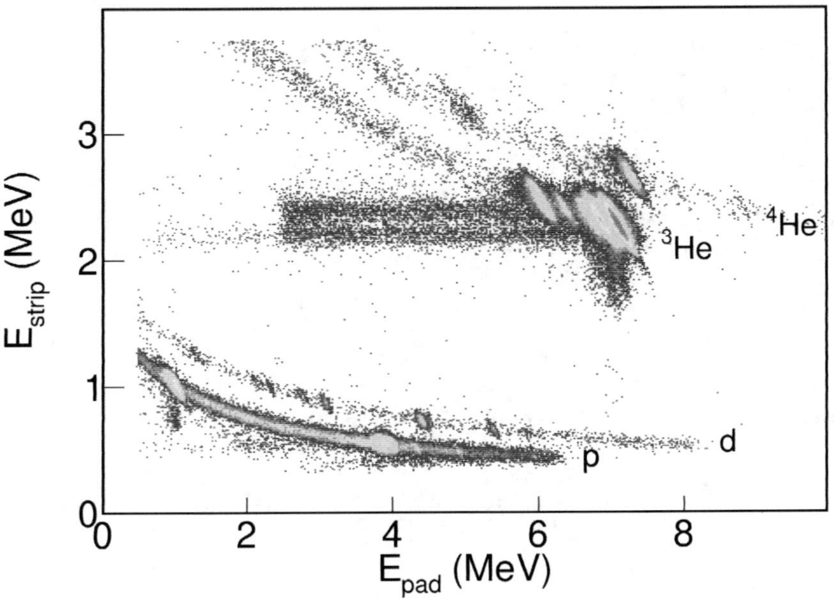

Figure 2. Experimental ΔE-E scatter plot for one of the pseudo-telescope of the OSCAR device. Both axes are calibrated in MeV. Different identification lines are identified by labels. Bumps observed on the deuteron line are produced by (^{3}He,d) on ^{32}S and the contaminants ^{12}C and ^{16}O.

quality of the energy calibration while merging data collected at different times during the experiment, we have produced a correction factor for each pseudo-telescope and each different run of the experiment. This procedure allows to maximize the statistics on the recorded spectra by maintaining an excellent energy resolution.

The particle identification is obtained by means of the ΔE-E identification technique, using the signals produced by the first (E_{strip}) and the second (E_{pad}) detection stages within a pseudo-telescope. Figure 2 shows a typical ΔE-E plot of a pseudo-telescope. Loci corresponding to protons, deuterons, ^{3}He and ^{4}He are clearly identified, as indicated by labels. We use a graphical cut procedure to unambiguously identify particles. Bumps lying on the observed ΔE-E loci are produced mainly by the occurrence of binary nuclear processes, such as scattering or binary reactions. Particularly interesting is the case of the deuteron locus, for which several bumps are well visible with a small background level, indicating the emission of deuterons from (^{3}He,d) transfer reactions. In the following analysis process, we will focus exclusively on well-identified deuterons.

In order to properly disentangle various (^{3}He,d) reactions induced on our target, we studied the kinematic lines associated to well-reconstructed deuterons in OSCAR. This is an advantage, with respect to previous investigations, introduced by the use of a new generation hodoscope, where many detection angles are achieved within the same geometry configuration. Figure 3 shows the deuteron kinetic energy as a function of the detection angle (θ) deduced with the procedure described above. The figure is produced by summing together data from 128 pseudo-telescopes, 64 for each of the two geometry configuration used during the experiment, covering

IOP Conf. Series: Journal of Physics: Conf. Series **1308** (2019) 012007 doi:10.1088/1742-6596/1308/1/012007

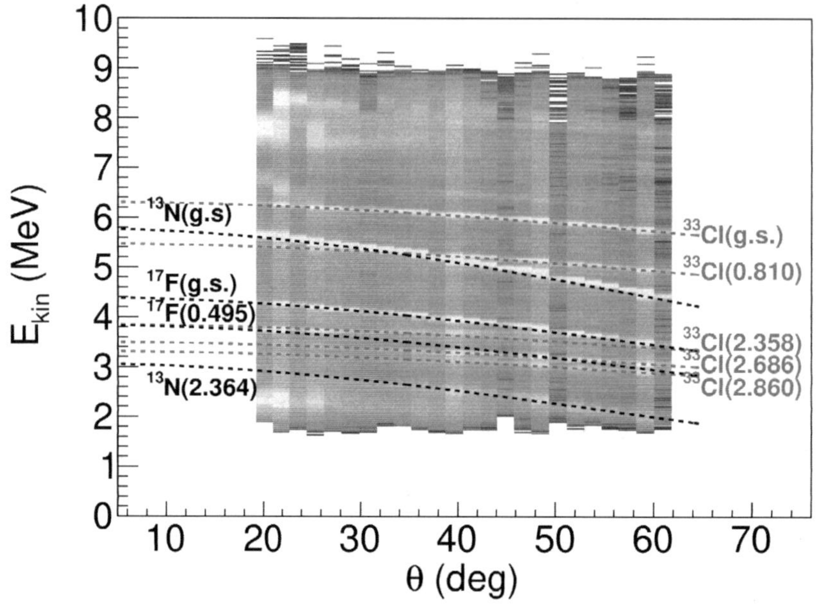

Figure 3. Experimental deuteron kinematic lines E_{kin} vs θ in the laboratory frame obtained with the OSCAR hodoscope and merging data from both the geometry configurations used in the experiment. Black and red dashed lines show the calculated kinematic lines respectively for contaminant reactions and ^{32}S(^{3}He,d)^{33}Cl reactions. ^{12}C and ^{16}O contaminants in the target are responsible for ^{13}N and ^{17}F residual nuclei lines, respectively.

a region of polar angles $20° \leq \theta \leq 60°$ in the laboratory frame. Several lines are clearly visible in the figure, indicating different (^{3}He,d) reactions. Red dashed lines are the result of our calculation assuming that the deuterons are emitted from the ^{32}S(^{3}He,d)^{33}Cl reactions, for different excitation energy states of the residual ^{33}Cl nucleus, indicated by labels. Black dashed lines are concurrent (^{3}He,d) reactions occurring on ^{12}C and ^{16}O contaminants present in the target backing. The latter give respectively ^{13}N and ^{17}F as the residual nuclei. In our calculation we take into account the energy loss by the beam in the target before reacting, assuming that the reaction occurs at mid target, and the energy loss by the emitted deuteron inside the target. Calculated lines are in excellent agreement with the observed ones, indicating the good consistency of our kinematics assumptions. In all the angular regions where the lines don't significantly overlap we can extract the yield of the ^{32}S(^{3}He,d)^{33}Cl reactions unambiguously.

The angular range shown in Figure 3 is segmented into angular regions according to the recorded statistics and to minimize the θ overlap between pseudo-telescopes of contiguous regions. We obtain over 30 separated angular regions. Within each region, we performed a multi-gaussian fit of the recorded peaks in the deuteron E_{kin} spectrum, according to the findings of Figure 3. In order to achieve a good consistency of the fit, we perform a simultaneous fit of data by using a function which contains the sum of all the gaussians. A typical example of such fits is shown in Figure 4, where the red line is the result of the fit. The fit reproduces well the overall shape of data. Peaks corresponding to the different reactions occurring on our ^{nat}Zn^{32}S target and the contaminant elements are identified by labels. The figure shows data

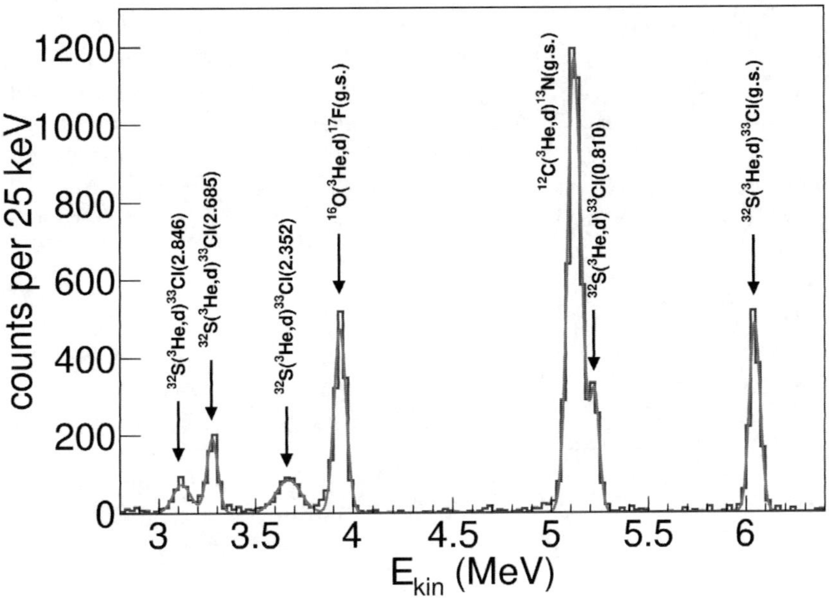

Figure 4. Experimental deuteron energy spectrum measured at $\theta \approx 40°$ with the OSCAR hodoscope. Data is obtained by merging 4 pseudo-telescopes. The red line is the result of a simultaneous fit of the spectrum with the superposition of gaussians to reproduce the observed peaks. Reactions responsible for each of the observed peaks are identified by labels.

within an angular slice centered at about $\theta = 40°$. The contribution of ^{32}S(^{3}He,d)^{33}Cl reactions leaving the residual ^{33}Cl in its ground state, the 2.352 MeV, the 2.685 MeV and the 2.846 MeV states are clearly visible, well separated from the contaminant peaks, as expected from the results shown in Figure 3. The ^{16}O(^{3}He,d)^{17}F(0.495) peak is strongly suppressed at this angle, and its contribution to the high energy tail of the ^{32}S(^{3}He,d)^{33}Cl(2.686) is negligible. The ^{32}S(^{3}He,d)^{33}Cl(0.810) is partially overlapped with the prominent ^{12}C(^{3}He,d)^{13}Cl(g.s.) peak, because of the kinematics, but, thanks to the excellent resolution of our apparatus, we are able to resolve both yields. The latter is particularly important at all angles, representing the major contaminant in our investigation, and making it extremely difficult to extract the yield of the ^{32}S(^{3}He,d)^{33}Cl(0.810) reaction in the angular region $25° \leq \theta \leq 35°$. However, at angles greater than $\approx 40°$ the separation in energy of these reactions is sufficiently large to allow an easy separation of the corresponding yields.

Finally, to extract the absolute cross-section of each of the measured ^{32}S(^{3}He,d)^{33}Cl reactions, we reduced indeterminations arising from the target thickness and the beam charge integration by internal normalization to the ^{3}He + Zn Rutherford scattering. The latter appears as a prominent peak in the ^{3}He kinetic energy spectra, well visible at all angles.

4. Conclusions and Perspectives

To try to solve the ambiguities in the experimental knowledge of low-lying states in the proton-rich ^{33}Cl nucleus, we have performed an investigation of the ^{32}S(^{3}He,d)^{33}Cl one-proton transfer reaction by using the new generation hodoscope OSCAR. The combination of the excellent energy resolution, the particle identification capabilities and the presence of several detection angles allow us to firmly extract the yield of ^{32}S(^{3}He,d)^{33}Cl reactions over the background caused by contaminants, in a wide angular domain $20° \leq \theta \leq 60°$. We are able to extract the yield of ^{32}S(^{3}He,d)^{33}Cl reactions leaving the ^{33}Cl residual nucleus in the ground state and in its 0.810 MeV, 2.352 MeV, 2.685 MeV, 2.846 MeV excited states. The absolute differential cross-sections will be deduced by normalization to the ^{3}He + Zn Rutherford scattering, thus removing any additional uncertainty related to the thickness of the target and the beam charge integration.

Our new data will be compared with the state-of-the-art shell model calculations used to describe nuclei in the (s-d)-shell to refine our knowledge of the spectroscopy of the observed states, including J^{π} values and the C^2S_p spectroscopic factors.

5. Acknowledgments

We acknowledge M. Loriggiola of the INFN-LNL Target Laboratory (Servizio Utenti) for manufacturing and characterizing extremely high quality and low contaminants targets. We are thankful to the staff of the CN accelerator of the INFN-LNL for producing and delivering the high-quality beams used for the experiment.

References

[1] Lombardo I *et al.* 2014 *Bulletin of the Russian Academy of Sciences: Physics* **78** 1093
[2] Lombardo I *et al.* 2014 *J. Phys. Conf. Ser.* **569** 012068
[3] Lombardo I, Dell'Aquila D, Spadaccini G, Verde G and Vigilante M 2018 *Phys. Rev. C* **97**(3) 034320 URL `https://link.aps.org/doi/10.1103/PhysRevC.97.034320`
[4] He J J *et al.* 2018 *Chin. Phys. C* **42** 15001
[5] Dell'Aquila D 2016 *Il Nuov. Cim. C* **39** 272
[6] Dell'Aquila D *et al.* 2017 *Phys. Rev. Lett.* **119** 132501
[7] Lombardo I *et al.* 2016 *J. Phys. G.* **43** 45109
[8] Spitaleri C *et al.* 2017 *Phys. Rev.* **C95** 035801
[9] Piantelli S *et al.* 2017 *Phys. Rev. C* **96** 034622
[10] Borderie B *et al.* 2018 *Phys. Lett. B* **782** 291
[11] DeFilippo E *et al.* 2009 *Acta. Phys. Pol. B* **40** 1199
[12] Morelli L *et al.* 2016 *J. Phys. G: Nucl. Part. Phys.* **43** 045110
[13] Russotto P *et al.* 2015 *Phys. Rev. C* **91** 014610
[14] Lombardo I *et al.* 2010 *Nucl. Phys. A* **834** 458c
[15] Pastore G *et al.* 2017 *Nucl. Instr. Meth. Phys. Res. A* **860** 42
[16] Lombardo I *et al.* 2013 *Nucl. Instr. Meth. Phys. Res. B* **302** 19
[17] Cappuzzello F *et al.* 2016 *Eur. Phys. J. A* **52** 167
[18] Fernández-Domínguez B *et al.* 2011 *Phys. Rev. C* **84** 011301
[19] Nevelinga R *et al.* 2011 *Nucl. Instr. Meth. Phys. Res. A* **654** 29
[20] Acosta L *et al.* 2016 *J. Phys. Conf. Ser.* **730** 012001
[21] Dell'Aquila D *et al.* 2019 *Nucl. Instr. Meth. Phys. Res. A* **929** 162
[22] Cardella G *et al.* 2015 *Nucl. Instr. Meth. Phys. Res. A* **799** 64
[23] Martorana N *et al.* 2018 *Phys. Lett. B* **782** 112
[24] Adsley P *et al.* 2017 *JINST* **12** T02004
[25] Dell'Aquila D *et al.* 2018 *Nucl. Instr. Meth. Phys. Res. A* **877** 227
[26] Wildenthal B *et al.* 1971 *Phys. Rev. C* **4** 1708
[27] Castel B *et al.* 1971 *Nucl. Phys. A* **162** 273
[28] Chen J and Sing B 2011 *Nucl. Data Sheets* **112** 1393
[29] Elbakr S *et al.* 1971 *Can. J. Phys.* **50** 674
[30] Kozub R and Youngblood D 1972 *Phys. Rev. C* **5** 413
[31] Inglima G *et al.* 1975 *Nuov. Cim. A* **26** 211

IOP Conf. Series: Journal of Physics: Conf. Series **1308** (2019) 012008 doi:10.1088/1742-6596/1308/1/012008

Pion induced Drell-Yan: the transverse momentum structure of the pion

A. Courtoy

Instituto de Física, Universidad Nacional Autónoma de México, Apartado Postal 20-364, 01000 Ciudad de México, México

E-mail: aurore@fisica.unam.mx

Abstract. We present an analysis of unpolarized Drell-Yan pair production in pion-nucleus scattering with a particular focus into the pion dynamics. The study consists in analyzing the effect of the partonic longitudinal and, especially, transverse distributions of the pion in a Nambu–Jona-Lasinio (NJL) framework, with Pauli-Villars regularization. In order to consistently take into account the QCD evolution effects, we have estimated the hadronic scale corresponding to the NJL model's degrees of freedom through a minimization procedure at NLO: The NLO evolved pion distributions have been compared to rapidity differential Drell-Yan cross sections data. That hadronic scale so determined represents the only free parameter in our approach.

The NJL transverse momentum PDF, evolved up to next-to-leading logarithmic accuracy, is then tested against the transverse momentum spectrum of dilepton pairs up to a transverse momentum of 2 GeV. We found a fair agreement with available pion-nucleus data. We find sizable evolution effects on the shape of the distributions and on the generated average transverse momentum of the dilepton pair.

1. Introduction

Drell-Yan (DY) processes have led to fascinating and challeging physics developments since the '60s. The process is described as follows

$$h_1(p_1)\, h_2(p_2) \to \gamma^*(q) + X, \tag{1}$$

in which a virtual photon is produced with large invariant mass-squared Q^2 in the collisions of two hadrons at a centre-of-mass energy $s = (p_1 + p_2)^2$, with $p_{1,2}$ the four momentum of hadrons $h_{1,2}$, respectively.

First of all, most fits of parton distribution functions (PDFs) rely on DY data. That is particularily true for pion PDFs whose few fits heavily depend on the pion-induced DY data [1] for which h_1 in Eq. (1) is a pion and h_2 a proton. The latter are the ones that we will consider in the present proceedings.

Beyond collinear approaches, the so-called unintegrated cross-sections characterize the spectrum of transverse momentum of the virtual photon, q_T. In the kinematical regime in which q_T is of order Λ_{QCD}, that is small *w.r.t.* Q, such an effect is accounted into transverse momentum of the partons through transverse-momentum dependent distributions (TMDs). The departure from collinearity is here a highly non-perturbative effect, originated in the internal dynamics of the parent hadron. TMDs have been studied for more than a decade now. While there exist model predictions as well as phenomenological analyses, the implementation of the transverse momentum factorization theorems has added in complexity in globally fitting and phenomenologically determining TMDs. In particular, Semi-Inclusive DIS has been the leading process for such studies, mainly due to experimental tendencies and successes. With the forthcoming pion-induced DY at COMPASS-II, the state-of-the-art theoretical framework will be tested and adjusted from unpolarized to polarized observables.[1] In that context, the structure of the pion plays an important rôle. In Ref. [2], we presented one of the first analyses of the πN DY process in terms of the modern TMD formulation. Our method focuses on investigating the DY cross section from the perspective of the dynamics of the pion as embodied by the Nambu–Jona-Lasinio (NJL) model [3]. Complementary studies have been performed in a Gaussian approach in Ref. [4] and a phenomenological fit of non-perturbative parameters of the TMD formulation is presented in Ref. [5].

On the other hand, in the complementary kinematical regime, $q_T \sim Q$, perturbative QCD corrections are expected to suffice to explain the transverse momentum spectrum in the regime in which it should apply. However, recent analyses of DY processes find that theoretical predictions based on fixed-order perturbation theory fail to describe Drell-Yan data [6]. This subject is beyond the scope of theses proceedings and our analysis, we however point out that it represents one of the biggest challenge for the next future.

These proceedings are organized as follows. In Section 2, the pion unpolarized TMD as obtained in the NJL model in Ref. [3] is described. Also, a new evaluation of the hadronic scale of the model is presented. In Section 3 we jump to the results with no further details about our choice for the proton structure.

2. Pion distribution in NJL

In this Section we synthetize the most important results of the the calculation of pion TMDs in a NJL framework, with Pauli-Villars regularization of Ref. [3]. Model calculations of meson partonic structure within this approach have a long story of successful predictions, see *e.g.* references in [7].

The unpolarized pion TMD is defined as

$$f^{q/\pi}(x_\pi, \mathbf{k}_T; Q_0^2) = \frac{1}{2}\int \frac{d\xi^- d^2\xi_T}{(2\pi)^3} e^{-i(\xi^- k^+ - \boldsymbol{\xi}_T \mathbf{k}_T)} \langle \pi^+ | \bar{\psi}(\xi^-, \boldsymbol{\xi}_T)\gamma^+ \frac{1+\tau_3}{2}\psi(0, \mathbf{0}_T)|\pi^+\rangle \quad . \tag{2}$$

The pion TMD obtained in the NJL framework, here in the chiral limit —which is an excellent approximation to the full result, is given by

$$f^{q/\pi}(x_\pi, \mathbf{k}_T; Q_0^2) = q(x_\pi; Q_0^2) T(\mathbf{k}_T) \ , \tag{3}$$

with $q(x_\pi; Q_0^2)$ the collinear pion PDF with momentum fraction x_π and Q_0 is the hadronic scale corresponding to the NJL model. In Eq. (3), it becomes appearant that the $(x, \mathbf{k}_T)$ dependences of the pion TMD factorize ata such a hadronic scale. Notice that it is no longer the case analytically in the full result. Also, we have assumed that the Q_0^2 dependence is carried by the collinear PDF exclusively ; it is not necessarily legitimate in QCD. In the chiral limit, one has (for π^-, of interest here) the well-known result

$$q(x_\pi, Q_0^2) = d_v(x_\pi, Q_0^2) = \bar{u}(x_\pi, Q_0^2) = 1 \, . \tag{4}$$

The function T is given by

$$T(\mathbf{k}_T) = \frac{3}{4\pi^3}\left(\frac{m}{f_\pi}\right)^2 \sum_{i=0,2} \frac{c_i}{k_T^2 + m_i^2} = T(k_T) \, , \tag{5}$$

[1] The main physics motivation of such an experiment is to crucially test the universality of the Sivers function, which is expected to have an opposite sign in DY *w.r.t* Semi-Inclusive DIS.

which satisfies the normalization

$$\int d^2\mathbf{k}_T \, T(\mathbf{k}_T) = 1 \,. \tag{6}$$

Among the required properties of the unpolarized TMD obtained in models, we stress that, upon integration over the quark intrinsic transverse momentum $\mathbf{k}_T$, the pion PDF $q(x)$ is properly recovered with correct normalization as demonstrate Eqs. (3,6), and the momentum sum rule is exactly satisfied. The recovery of these properties is due to the fact that NJL is a field theoretical scheme. Also, the correct support of the PDF, $0 \leq x \leq 1$, arises naturally here. Since the result Eq. (3) directly follows from the definition Eq. (2), the $\mathbf{k}_T$ dependence is automatically generated by the NJL dynamics. This is an important feature of the results of Ref. [3], not found in other approaches, *e.g.* [4, 5]. Since the distribution Eq. (5) depends only upon k_T^2, its Fourier transform can be cast in the form

$$\begin{aligned} S^{\pi}_{NP}(b) &= \frac{3}{2\pi^2}\left(\frac{m}{f_\pi}\right)^2 \sum_{i=0,2} \int dk_T \, k_T \, J_0(bk_T) \frac{c_i}{k_T^2 + m_i^2} \\ &= \frac{3}{2\pi^2}\left(\frac{m}{f_\pi}\right)^2 \sum_{i=0,2} c_i K_0(m_i \, b) \,, \end{aligned} \tag{7}$$

where K_0 is the modified Bessel function of the second kind.

In the above definition and results, an additional fundamental parameter is made explicit, *i.e.* the hadronic scale of the model Q_0^2. Distribution functions as evaluated in models must be matched to the particular low RGE scale at which they mimic the true theory the best. Here, the collinear parton distribution obtained within NJL are associated to a specific low momentum scale Q_0^2 and, in order to be used to predict measured quantities, have to be evolved to higher momentum scales according to perturbative QCD. Such a low scale has been determined previously by directly comparing the second moment of the pion PDF evaluated in NJL model with the results from the analysis of Ref. [8]. The procedure gives a value of $Q_0^2 = 0.18$ GeV^2 at NLO [7, 9][2].

In our analysis, we have used a different strategy : we consider Q_0^2 a free parameter of the NJL model which is then fixed with a minimization procedure of the theoretical $\pi^- W$ DY cross sections, differential in $\sqrt{\tau}$ and x_F, against the corresponding experimental data [1]. Theoretical cross sections are calculated according to

$$\frac{d^2\sigma}{dQ^2 dx_F} = \frac{4\pi\alpha_{em}^2}{9Q^2 s} \sum_{ij} e_i^2 \int_{x_1}^{1} dt_1 \int_{x_2}^{1} dt_2 \frac{d^2\hat{\sigma}^{ij}}{dQ^2 dx_F} f_{i/\pi}(t_1, Q^2) f_{j/p}(t_2, Q^2) \,, \tag{8}$$

where the partonic cross sections $d\hat{\sigma}^{ij}$ are calculated at NLO accuracy by using the results of Ref. [8]. The NJL pion PDFs are evolved to NLO accuracy in the Variable Flavor Number Scheme, with the initial condition given in Eq. (4), with the help of the QCDNUM [10] evolution code. The QCD parameters are those of the NLO CTEQ6M parameterisation [11]. In particular we set the NLO running coupling to $\alpha_s^{(n_f=5)}(M_Z) = 0.118$ at the Z-boson mass, M_Z. Since the data we are comparing to are obtained on a tungsten target, we take into account nuclear effects by using nuclear PDFs of Ref. [12]. We have carried out a χ^2 study to establish the hadronic scale of the model that describes the data the best at NLO in pQCD. Two cases have been considered: an evaluation of the χ^2 for the full range of x_F and another one with a cut $x_F < 0.4$,

[2] Other schemes give higher values for the hadronic scale, *i.e.* up to ~ 1 GeV^2 [13].

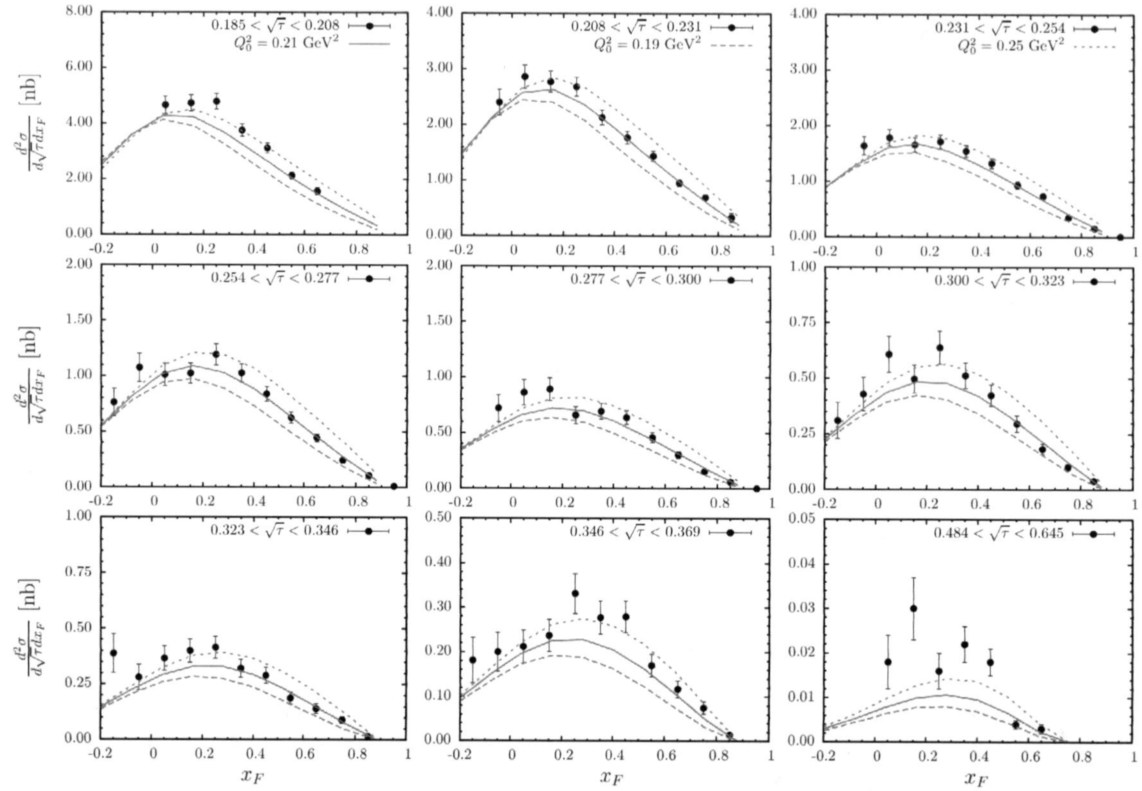

Figure 1. Drell-Yan pairs production in $\pi^- W$ collisions. Next-to-leading order cross sections obtained by using evolved NJL pion PDFs for three values of Q_0^2 are compared to data of Ref. [1].

since the NJL model is expected to better reproduce the pion valence distributions, expected to populate the range of large and positive x_F . The scales thus determined are

$$Q^2_{0,\,\text{no cut}} = 0.212^{+0.011}_{-0.012}\ \text{GeV}^2, \qquad Q^2_{0,\,\text{cut}} = 0.209^{+0.008}_{-0.009}\ \text{GeV}^2, \tag{9}$$

and correspond to a chisquare value of χ^2/d.o.f.= 2.1 and 1.9, respectively. The quoted errors correspond to a variation of one unit in χ^2, *i.e.* 1-σ. Those results are compatible with each other. We will therefore refer to $Q_0^2 = 0.21$ GeV^2, as the scale associated to the pion NJL model. In Fig. 1, three values for the hadronic scale are depicted for comparison. It is worth noticing that the results show an acceptable agreement, both in shape and in normalization.

3. Pion dynamics in DY cross sections

As mentioned earlier, when q_T^2 becomes small compared to Q^2, large logarithmic corrections of the form of $\alpha_s^n \log^m(Q^2/q_T^2)$ with $0 \le m \le 2n-1$ appear in fixed order results, being n the order of the perturbative calculation. These large logarithmic corrections can be resummed to all orders by using the Collins-Soper-Sterman (CSS) formalism [14]. In this limit, the cross-section,

IOP Conf. Series: Journal of Physics: Conf. Series **1308** (2019) 012008 doi:10.1088/1742-6596/1308/1/012008

differential in $\mathbf{q}_T$, can be written as

$$\frac{d\sigma}{dq_T^2 d\tau dy} = \sum_{a,b} \sigma_{q\bar{q}}^{(LO)} \int_0^\infty db \frac{b}{2} J_0(b\, q_T)\, S_q(Q,b)\, S_{NP}^{\pi p}(b)$$
$$\left[(f_{a/\pi} \otimes C_{qa})\left(x_1, \frac{b_0^2}{b^2}\right)(f_{b/p} \otimes C_{\bar{q}b})\left(x_2, \frac{b_0^2}{b^2}\right) + q \leftrightarrow \bar{q}\right], \quad (10)$$

where $b_0 = 2e^{-\gamma_e}$, the symbol $\otimes$ stands for convolution and $\sigma_{q\bar{q}}^{(LO)}$ is the leading-order total partonic cross section for producing a lepton pair. The a, b indices run on quark and gluons, and $J_0(b\, q_T)$ is the Bessel function of first kind. The cross section in Eq. (10) is also differential in $\tau = Q^2/s$ and y, the rapidity of the DY pair. Momentum fractions appearing in parton distribution functions can be expressed in terms of these variables as

$$x_{1(2)} = \sqrt{\tau} e^{\pm y}, \quad y = \frac{1}{2} \ln \frac{x_1}{x_2}. \quad (11)$$

Cross sections differential in $x_F = x_1 - x_2 = 2q_\parallel/\sqrt{s}$, the longitudinal momentum of the pair in the hadronic centre of mass system, can be obtained from those differential in rapidity y, *i.e.* $dy = dx_F/\sqrt{x_F^2 + 4\tau}$; we also have that $|x_F| < 1 - \tau$. The large logarithmic corrections, exponentiated in b-space in the Sudakov perturbative form factor, are expressed as

$$S_q(Q,b) = \exp\left\{-\int_{b_0^2/b^2}^{Q^2} \frac{dq^2}{q^2}\left[A(\alpha_s(q^2)) \ln \frac{Q^2}{q^2} + B(\alpha_s(q^2))\right]\right\}. \quad (12)$$

The functions C_{ab} and A, B have perturbative expansions in α_s and can be found *e.g.* in the original paper (see Ref. [2]).

On the other hand, the non-perturbative factor, $S_{NP}^{\pi p}$, contains all the information about the non-perturbative $\mathbf{k}_T$ behavior. More recent formulations of the CSS formalim directly include the TMD parton densities [15]. Models for hadron's structure yet, as mentioned above, might represent the true theory but at one specific value of the RGE scale, Q_0. In that sense, they incorporate a $\mathbf{k}_T$ —or equivalently its Fourier conjugate, as well as the Bjorken x, behavior in a fashion that can be resumed as,

$$f^{q/\pi}(x_\pi, b; Q_0^2) = q(x_\pi; Q_0^2) S_{NP}^{\pi}(b) = q(x_\pi; Q_0^2) \exp\{\ln S_{NP}^{\pi}(b)\} \; ; \quad (13)$$

while the full TMD parton densities should involve a further Q^2 dependence, that is also called *non-perturbative* evolution. Such an evolution has been parameterized in the past, for the proton-proton DY, as

$$S_{NP}^{pp}(b) = \exp\{-[a_1 + a_2 \ln(M/(3.2\,\text{GeV})) + a_3 \ln(100 x_1 x_2)]b^2\}, \quad (14)$$

where a_1 plays an equivalent rôle to $\ln S_{NP}^{\pi}(b)$ in Eq. (13) and the other parameters, determined *e.g.* in Ref. [16], reflect a Q^2 evolution, *i.e.* through the invariant mass M dependence, as well as an unfactorized $(x, \mathbf{k}_T)$ term.

Our ansatz here is therefore

$$S_{NP}^{\pi p}(b) = S_{NP}^{\pi}(b) \sqrt{S_{NP}^{pp}(b)}, \quad (15)$$

where $S_{NP}^{\pi}(b)$ is given in Eq. (7) and the square root on $S_{NP}^{pp}(b)$, given in Eq. (14).

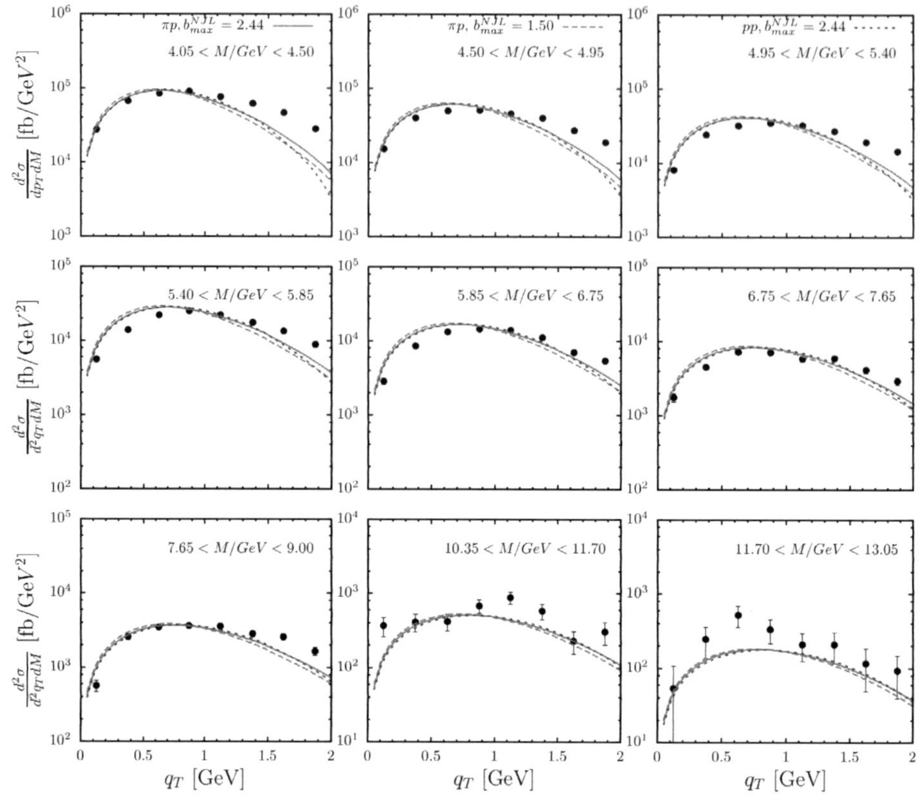

Figure 2. Our model results compared to cross sections in various invariant mass bins of the DY lepton pair integrated in $0 < x_F < 1$. Data from Ref. [1].

At very large values of b, the perturbative form factor needs to include a *taming* through the so-called $b^\star$-prescription. For the process of interest here, it is useful to split the perturbative form factor in Eq. (12) in a form which allows to use distinct b_{max} on the proton and pion side:

$$\begin{aligned} S_q(Q,b) &\equiv S_q(Q,b_\star,b^p_{max},b^\pi_{max}) \\ &= \exp\left\{-\int_{\frac{b_0^2}{b_\star^2(b^p_{max})}}^{Q^2}\frac{dq^2}{2\,q^2}\left[A(\alpha_s(q^2))\,\ln\frac{Q^2}{q^2}+B(\alpha_s(q^2))\right]\right\} \\ &\times \exp\left\{-\int_{\frac{b_0^2}{b_\star^2(b^\pi_{max})}}^{Q^2}\frac{dq^2}{2\,q^2}\left[A(\alpha_s(q^2))\,\ln\frac{Q^2}{q^2}+B(\alpha_s(q^2))\right]\right\}, \end{aligned} \tag{16}$$

with $b_\star(b,b_{max}) = b/\sqrt{1+\left(\frac{b}{b_{max}}\right)^2}$ and the respective $b^p_{max} = 1.5\text{GeV}^{-1}$ and the value of b^π_{max} is adopted such that $b^\pi_{max} = b_0/Q_0 = 2.44\text{GeV}^{-1}$.

In Fig. 2, we show the results for the lepton pair q_T-spectra of Ref. [1], measured in πW collisions. The cross sections differential in q_T are integrated over $0 < x_F < 1$ for various M values. The plots range up to $q_T \sim 2$ GeV, range for which the chosen proton description holds [2]. Both red, full and dashed, curves correspond to results using Eq. (15) with, respectively, the proposed regulator value $b^\pi_{max} = 2.44$ GeV^{-1} and $b^\pi_{max} = 1.5$ GeV^{-1}

IOP Conf. Series: Journal of Physics: Conf. Series **1308** (2019) 012008 doi:10.1088/1742-6596/1308/1/012008

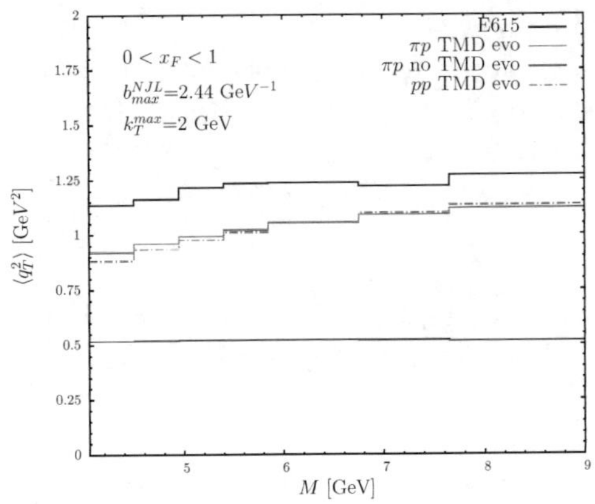

Figure 3. Lepton pair average transverse momentum, $\langle q_T^2 \rangle$, as a function of M integrated in the range $0 < x_F < 1$. Averaged values are obtained integrating both predictions and the phenomenological parametrization of the data up to $q_T^{max} = 2$ GeV.

demonstrating the stability of our results at small values of q_T. The short-dashed blue curve corresponds to a different ansatz: Eq. (14) is used for both hadrons, still with different b_{max} values. At low-q_T, the difference between the two ansätze is quantitatively small. It supports the hypothesis that the perturbative evolution, mainly driven by Eq. (16), washes away differences in the non-perturbative structure, *i.e.* gaussian-like with Q^2 dependence vs. dynamically generated $\mathbf{k}_T$ dependence at the hadronic scale Q_0. We are tempted to conclude that there is a reduced sensitivity of the data to non-perturbative structure. However, this is a first analysis for which no *non-perturbative evolution* of the type Eq. (14) has been included, since it is not inherent to the NJL approach used here. A similar conclusion can be driven for our results for the cross-sections differential in q_T but integrated over $4 < M < 8.55$GeV, in the range of x_F where the proton description of [16] is valid. For higher values of x_F, further analyses need to be done before drawing any conclusion.

The average transverse momentum of the pair, $\langle q_T^2 \rangle$, can also be evaluated, as a function of either x_F or M. The latter is shown in Fig. 3. It can be appreciated that the theoretical results undershoot the experiment's, yet reproducing its trend. A slight difference between the Gaussian+ Q^2-dependence (dot-dashed red curve) and the NJL pion distribution (full red curve) is seen at smaller values of the invariant mass of the lepton pair.

4. Conclusions

In these proceedings we synthetized the analysis of the DY pair production in pion-nucleus scattering presented in Ref. [2]. In that work we tested the outcome of a Nambu–Jona-Lasinio approach for the pion transverse momentum distribution plugged in the CSS framework at next-to-leading logarithmic accuracy against the differential transverse momentum spectra of DY pairs produced in pA collisions.

The Nambu–Jona-Lasinio approach is a low-energy model for the pion structure, to which is allocated a low hadronic scale, Q_0. That hadronic scale is the only free parameter of the model and, in our analysis, it has been determined again, with an adapted strategy, confirming its low value, $Q_0^2 = 0.21\text{GeV}^2$. The agreement found between our pion-nucleus theoretical cross sections and experimental data is fair, confirming the predictive power of the NJL model, for both the longitudinal pion parton distributions and its transverse structure, especially at low values of q_T where no additional contribution to the differential cross-section as described by the CSS formalism, Eq. (10), is needed.

Further analyses could include a customized extension of the current approach to include a Q^2 dependence. The upcoming data from COMPASS-II on unpolarized target need being understood before turning theorists' attention to the long awaited DY with transversely polarized proton target and its promising access to the Sivers function.

Acknowledgments

The author thanks her colleagues and co-authors of the original publication, F. A. Ceccopieri, S. Noguera and S. Scopetta. This work is supported in part by DGAPA-PAPIIT IA102418.

References

[1] J. S. Conway *et al.*, Phys. Rev. D **39** (1989) 92. doi:10.1103/PhysRevD.39.92
[2] F. A. Ceccopieri, A. Courtoy, S. Noguera and S. Scopetta, Eur. Phys. J. C **78** (2018) no.8, 644 doi:10.1140/epjc/s10052-018-6115-3 [arXiv:1801.07682 [hep-ph]].
[3] S. Noguera and S. Scopetta, JHEP **1511** (2015) 102 doi:10.1007/JHEP11(2015)102 [arXiv:1508.01061 [hep-ph]].
[4] B. Pasquini and P. Schweitzer, Phys. Rev. D **90** (2014) no.1, 014050 doi:10.1103/PhysRevD.90.014050 [arXiv:1406.2056 [hep-ph]].
[5] X. Wang, Z. Lu and I. Schmidt, JHEP **1708** (2017) 137 doi:10.1007/JHEP08(2017)137 [arXiv:1707.05207 [hep-ph]].
[6] A. Bacchetta, G. Bozzi, M. Lambertsen, F. Piacenza, J. Steiglechner and W. Vogelsang, arXiv:1901.06916 [hep-ph].
[7] A. Courtoy, Ph. D. Thesis, Valencia University, 2009. http://arxiv.org/abs/arXiv:1010.2974
[8] P. J. Sutton, A. D. Martin, R. G. Roberts and W. J. Stirling, Phys. Rev. D **45** (1992) 2349. doi:10.1103/PhysRevD.45.2349
[9] A. Courtoy and S. Noguera, Phys. Lett. B **675** (2009) 38 doi:10.1016/j.physletb.2009.03.070 [arXiv:0811.0550 [hep-ph]].
[10] M. Botje, Comput. Phys. Commun. **182** (2011) 490 doi:10.1016/j.cpc.2010.10.020 [arXiv:1005.1481 [hep-ph]].
[11] J. Pumplin, D. R. Stump, J. Huston, H. L. Lai, P. M. Nadolsky and W. K. Tung, JHEP **0207** (2002) 012 doi:10.1088/1126-6708/2002/07/012 [hep-ph/0201195].
[12] K. Kovarik *et al.*, Phys. Rev. D **93** (2016) no.8, 085037 doi:10.1103/PhysRevD.93.085037 [arXiv:1509.00792 [hep-ph]].
[13] S. Noguera and V. Vento, Eur. Phys. J. A **28** (2006) 227 doi:10.1140/epja/i2006-10045-5 [hep-ph/0505102].
[14] J. C. Collins, D. E. Soper and G. F. Sterman, Nucl. Phys. B **250** (1985) 199. doi:10.1016/0550-3213(85)90479-1
[15] J. Collins and T. Rogers, Phys. Rev. D **91** (2015) no.7, 074020 doi:10.1103/PhysRevD.91.074020 [arXiv:1412.3820 [hep-ph]].
[16] A. V. Konychev and P. M. Nadolsky, Phys. Lett. B **633** (2006) 710 doi:10.1016/j.physletb.2005.12.063 [hep-ph/0506225].

IOP Conf. Series: Journal of Physics: Conf. Series **1308** (2019) 012009 doi:10.1088/1742-6596/1308/1/012009

Nuclear Astrophysics deep underground and the LUNA experiment

Rosanna Depalo

INFN - Sezione di Padova, Via Marzolo 8, 35131 Padova, Italy

E-mail: rdepalo@pd.infn.it

Abstract. The cross sections of nuclear reactions relevant for astrophysics are crucial ingredients to understand the energy generation inside stars and the synthesis of the elements. In stars, nuclear reactions take place at energies well below the Coulomb barrier. As a result, their cross sections are often too small to be measured in laboratories on the Earth's surface, where the signal would be overwhelmed by the cosmic-ray induced background. An effective way to suppress the cosmic-ray induced background is to perform experiments in underground laboratories. LUNA is a unique facility located at Gran Sasso National Laboratories (Italy) and devoted to Nuclear Astrophysics. The extremely low background achieved at LUNA allows to measure nuclear cross sections directly at the energies of astrophysical interest. Over the years, many crucial reactions involved in stellar hydrogen burning as well as Big Bang Nucleosynthesis have been measured at LUNA. This paper provides a short overview on underground Nuclear Astrophysics and discusses the latest results and future perspectives of the LUNA experiment.

1. Introduction

Nuclear fusion cross sections are relevant to many fields of astrophysics, including primordial nucleosynthesis, stellar evolution and nucleosynthesis, solar neutrino, gamma-ray astronomy etc. The whole life of stars can be described as a sequence of phases in which heavier and heavier elements are burnt inside the stellar core. Starting from the primordial hydrogen and helium, a variety of elements is produced at each step and theoretical models try to match predicted elemental abundances with astronomical observations.

With the advent of precision astronomy, the abundances of the elements in the outer layers of stars are now measured with precisions at the percent level, calling for increasingly-accurate inputs to the stellar models. A much-needed input in order to predict stellar nucleosynthesis is the thermonuclear reaction rate (i.e. the number of reactions per unit time and volume occurring in a star at a given temperature) of all the nuclear reactions involved.

At typical astrophysical temperatures, the kinetic energy of the interacting particles is usually much lower than the Coulomb repulsive potential between the positively-charged nuclei. Therefore, nuclear reactions can only occur by quantum mechanical tunneling and the cross section decreases steeply with the energy [1]. Because of the interplay between the Maxwell-Boltzmann energy distribution of nuclei and the tunneling probability through the Coulomb barrier, thermonuclear reactions can only occur in a well-defined energy range, called the Gamow peak. At Gamow energies, nuclear cross sections can become extremely small (of the order of 10^{-9} - 10^{-12} barn), therefore, in typical experimental conditions, the expected counting rate can be much smaller than the environmental background in a detector.

IOP Conf. Series: Journal of Physics: Conf. Series **1308** (2019) 012009 doi:10.1088/1742-6596/1308/1/012009

2. Underground Nuclear Astrophysics

Environmental background is produced by cosmic rays and the decay of naturally-occurring radioactive isotopes (uranium and thorium chains and ^{40}K). The background from radioactive isotopes can be substantially reduced by shielding the detector with high-Z and high-density material (usually lead or copper). On the other hand, cosmic radiation at sea level is mainly made of muons, highly penetrating particles created in the upper atmosphere that leave a trace in most particle and gamma ray detectors and can generate spallation reactions in detector and surrounding materials, with consequent production of neutrons and radioactive nuclei.

An effective way to suppress environmental background is to perform experiments in deep-underground laboratories. Indeed, not only can the cosmic muon flux be substantially suppressed by a thick rock overburden, but detectors can also be shielded from low-energy gamma background using much thicker linings than overground because secondary emission of radiation from the interaction of cosmic rays within the shielding becomes negligible [2, 3, 4].

LUNA (Laboratory for Underground Nuclear Astrophysics), was the pioneering experiment in deep underground Nuclear Astrophysics, being operative since 1992. More recently, other deep-underground [5, 6] and shallow-underground [7] experiments devoted to Nuclear Astrophysics have been developed worldwide.

In the following section, latest results and future perspectives of the LUNA experiment will be outlined.

3. The LUNA experiment

LUNA is located at the Gran Sasso underground laboratories (LNGS), in central Italy. The laboratory is shielded by 1400 meters of mountain rocks, attenuating the cosmic-ray muon flux by about six orders of magnitude when compared with the surface of the Earth. Fig. 1 shows a comparison of HPGe background spectra taken overground and at LNGS with two different types of passive shielding.

The LUNA laboratory was started in 1992 with the installation of a 50-kV accelerator [9]. In 2001 the 50-kV machine was replaced by a 400-kV accelerator, which is still operational today. A third facility with a more energetic accelerator, named LUNA MV, will be installed in 2020. The LUNA 400-kV accelerator provides proton and alpha-particle beams with intensity as high as 500 μA on target. The beam energy stability is particularly important for nuclear astrophysics experiments, since the fusion cross section below the Coulomb barrier depends exponentially on the energy. The LUNA proton beam energy is calibrated with 0.3-keV accuracy and has a long term stability of 5 eV/h. The beam energy spread has been measured to be 100 eV [10]. Two beam lines are available at LUNA: one equipped with a solid-target setup and the other with a windowless gas target. Different gamma-ray or particle detectors can be used, tuning the detection system according to the specific needs of the nuclear reaction to be studied.

The LUNA setup is ideally suited for the study of hydrostatic hydrogen burning taking place in stars at temperatures of 0.01 - 0.1 GK [11] and also explosive burning at temperatures up to 1 GK in sites such as the Big Bang [12], classical novae and supernovae [13].

Over the years, a large number of reactions involved in stellar hydrogen burning and Big Bang Nucleosynthesis have been investigated at LUNA. The cross section of those reactions was measured either within or very close to the Gamow window [14, 15].

4. Latest results and ongoing measurements

4.1. The ^{17}O(p,α)^{14}N reaction

Proton capture on ^{17}O is part of the CNO cycle of hydrogen burning and it proceeds through two alternative channels: ^{17}O(p,γ)^{18}F and ^{17}O(p,α)^{14}. The ratio of the (p,γ) to (p,α) cross sections determines the leak rate from the second to the third CNO cycle (see Fig. 2) and it affects the isotopic abundances of ^{17,18}O and ^{18,19}F and the flux of 511 keV gamma rays from

IOP Conf. Series: Journal of Physics: Conf. Series **1308** (2019) 012009 doi:10.1088/1742-6596/1308/1/012009

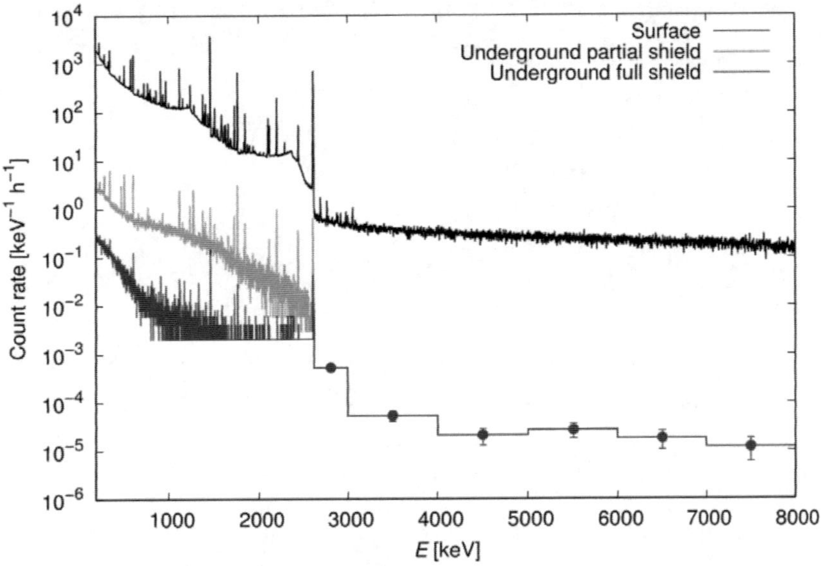

Figure 1. Environmental background spectra measured with an HPGe detector positioned at the Earth's surface, at LUNA with full passive shielding (4 cm thick copper, 25 cm thick lead and a radon box [8]) and at LUNA with a partial passive shield (4 cm thick copper and 25 cm thick lead) designed to host a second HPGe detector [32]. Above 2600 keV, the backgrounds in the two shielded cofigurations coincide.

classical novae, due to β^+ decay of ^{18}F.
At astrophysical energies, the ^{17}O(p,γ)^{18}F cross section depends on the narrow resonances at 64.5 keV and 183 keV, lying on the tail of two broad resonances at 557 keV and 677 keV, and on the direct capture component. The 183-keV resonance and the direct-capture cross section have both been studied at LUNA combining prompt gamma-ray spectroscopy and the activation technique [16].
The ^{17}O(p,α)^{14}N reaction has also been investigated at LUNA. Its cross section is also dominated by the resonances at 183 keV and 64.5 keV, but while the 183-keV resonance has a well known strength, the 64.5-keV resonance was debated: the first direct measurement performed gave an upper limit on the strength $\omega\gamma < 0.8$ neV [17], later revised to $\omega\gamma < 22$ neV [18]. A second direct experiment provided a strength of $\omega\gamma = 5.5^{+1.8}_{-1.5}$ neV [19] (adopted by the NACRE compilation [20]), later re-analyzed multiple times to give $\omega\gamma = 4.8 \pm 0.5$ neV (adopted in the reaction rate compilation by C. Iliadis et al. [21]). Finally, indirect measurements performed with the Trojan-horse technique provided $\omega\gamma = 3.4 \pm 0.6$ neV [22].
At LUNA, the ^{17}O(p,α)^{14}N reaction has been studied delivering the proton beam to solid Ta_2O_5 targets enriched in ^{17}O. Alpha particles from the reaction were detected with an array of eight Si detectors surrounded by a 5 cm thick lead shielding [23]. The natural background observed in the detectors was a factor of 15 lower compared to overground [23]. Thanks to this, both the 183-keV resonance and the weak resonance at 64.5 keV could be properly measured by LUNA in ^{17}O(p,α)^{14}N [24].
For the 183-keV resonance an $\omega\gamma = (1.68 \pm 0.03_{stat} \pm 0.12_{syst})$ meV was found, in good agreement with the literature.
For the 64.5-keV resonance, a strength $\omega\gamma = (10.0 \pm 1.4_{stat} \pm 0.7_{syst})$ neV was measured [24]. Such a value is about a factor of 2 higher than the previously estimated, thus leading to a factor

of 2 increase in the reaction rate of shell hydrogen burning in red giant and AGB stars (see Fig. 2).

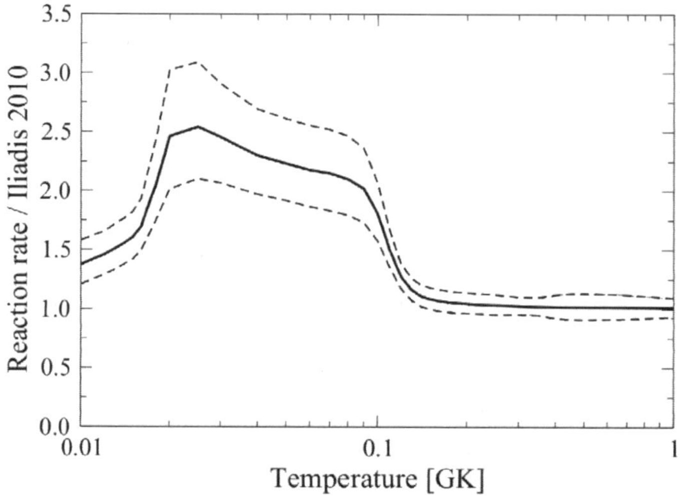

Figure 2. $^{17}O(p,\alpha)^{14}N$ reaction rate obtained adopting the LUNA strength for the 64.5 keV resonance

The new resonance strength has increased by 20% the predicted $^{16}O/^{17}O$ ratio after the first dredge up in red giant stars [25] and it has allowed for the firm identification of the production site of a oxygen-rich star grain population [26].

4.2. The $^{22}Ne(p,\gamma)^{23}Na$ reaction

The neon-sodium cycle of hydrogen burning contributes to the synthesis of the isotopes between ^{20}Ne and ^{23}Na in Red Giant Branch (RGB) stars [27], Asymptotic Giant Branch (AGB) stars [28] and classical novae explosions [29]. The $^{22}Ne(p,\gamma)^{23}Na$ reaction rates reported in the widely adopted compilations by NACRE [20] and C. Iliadis et al. [21] are three orders of magnitude discrepant in the energy range of interest for hydrogen burning in AGB stars.

The $^{22}Ne(p,\gamma)^{23}Na$ Gamow window for AGB stars and classical novae extends from 50 to 600 keV. In this energy range, the proton capture on ^{22}Ne is dominated by a large number of resonances. While some of the resonances could be measured with good precision in overground experiments [30, 31], the resonances below 400 keV, including two tentative resonances at 70 and 100 keV, and the direct capture contribution were the subject of an extensive experimental campaign at LUNA. The two tentative resonances are actually the reason why the rates from the two compilations are so different. Indeed, while NACRE decided to keep the two resonances with their upper limits, Iliadis et al. decided to disregard them.

The LUNA measurements were organized in two campaigns, both exploiting the windowless gas target system filled with 99.9% enriched ^{22}Ne: in the first campaign, gamma rays were detected using two high-resolution HPGe detectors [32, 33, 34] while the second campaign made use of a high-efficiency BGO detector [35]. In the first campaign the resonances at 156.2, 189.5 and 259.7 keV were observed for the first time in a direct experiment. For these resonances, the complete excitation function was measured and then a long run at the energy of maximum yield was performed. New gamma-decay modes have also been observed for the three resonances detected. For the non-detected resonances new upper limits have been measured. The new upper limits are two to three orders of magnitude lower than the previous direct measurement, proving the improvement in sensitivity that can be achieved in underground experiments [33, 34].

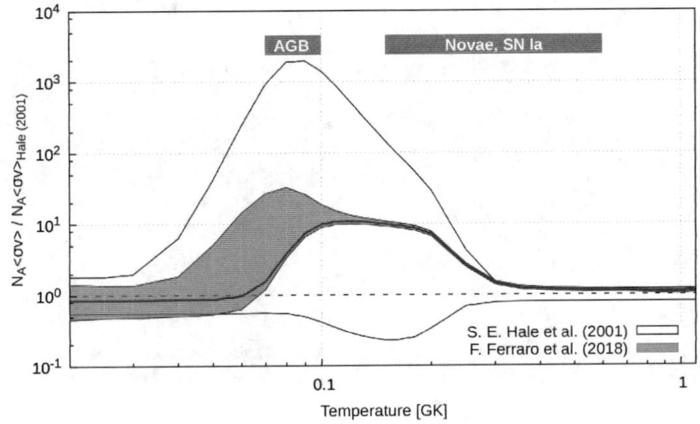

Figure 3. Reaction rate calculated adopting the LUNA results on the ^{22}Ne(p,γ)^{23}Na reaction

The upper limits on the resonances at 70 and 100 keV were further reduced during the BGO phase and the non-resonant contribution to the cross section was measured for the first time in the energy range between 100 and 300 keV in the center of mass. Fig. 3 shows the updated reaction rate for the ^{22}Ne(p,γ)^{23}Na reaction, calculated including the new LUNA results. The reaction rate adopted in [36] is also shown for comparison. The two tentative resonances at 70 and 100 keV are included in the calculation of the median and high rates, while the low rate is calculated disregarding the two resonances.

4.3. Ongoing experiments

A rich scientific program is underway at LUNA. Experiments for which the data analysis was just finalized include the study of the ^{23}Na(p,γ)^{24}Mg reaction with both a germanium and a BGO detector, and of the ^{18}O(p,γ)^{19}F and ^{18}O(p,α)^{15}N [37] reactions.
Two reactions for which the data taking was just concluded are ^{2}H(p,γ)^{3}He and ^{6}Li(p,γ)^{7}Be, briefly described below.

The ^{2}H(p,γ)^{3}He reaction is involved in the Big Bang Nucleosynthesis (BBN) network and it determines the abundance of primordial deuterium. Observational information on the primordial deuterium abundance is obtained with percent precision from the spectroscopic measurement of metal-poor damped Lyman-alpha (DLA) systems. These systems are the oldest astrophysical environments where deuterium is detected. The most recent observed deuterium abundance is $[D/H] = (2.527 \pm 0.030) \cdot 10^{-5}$ [38], while predictions oscillate from $(2.439 \pm 0.052) \cdot 10^{-5}$ to $(2.587 \pm 0.055) \cdot 10^{-5}$ depending on the cross section adopted for the ^{2}H(p,γ)^{3}He reaction [39]. As a matter of fact, only a few experimental points exist for the ^{2}H(p,γ)^{3}He cross section in the energy region of interest, with an overall systematic uncertainty at the 6-10% level [40].
The cross section of the ^{2}H(p,γ)^{3}He reaction was already measured at LUNA with the 50-kV accelerator to explore the energy region where the reaction takes place in pre-main sequence stars and in the Sun. A new experiment has now been performed at the 400-kV accelerator to cover the energy range of BBN. The experiment was performed using the windowless gas target filled with deuterium at a pressure of 0.3 mbar, surrounded by two different types of gamma detectors, each in a different phase of the data taking: a 4π BGO and one HPGe detector (see

fig. 4). The geometry of the BGO reduces the dependence of the detector response on the angular distribution of the emitted gamma rays [41]. On the other hand, the second phase of the experiment allows to infer the angular distribution exploiting the high energy resolution of the germanium detector and the Doppler shift of gamma ray energies.

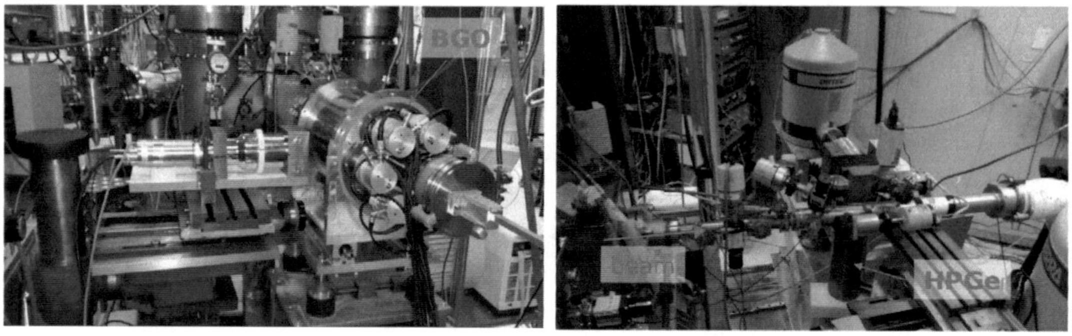

Figure 4. Photos of the two experimental setups adopted for the study of the ^{2}H(p,γ)^{3}He reaction

The goal of the experiment is to reach an overall systematic uncertainty on the 3% level combining the results from the two phases.

The ^{6}Li(p,γ)^{7}Be and ^{6}Li(p,α)^{3}He reactions are responsible for lithium-6 depletion in the early stages of stellar evolution and contribute to Bang Nucleosynthesis. The ^{6}Li(p,α)^{3}He reaction has been studied by many groups, and in order to explain the angular distribution of the emitted alpha particles, an R-matrix fit of the experimental data requires the contribution of both negative- and positive-parity excited states [42]. The existence of positive-parity excited states in ^{7}Be has never been verified experimentally. A recent measurement of the ^{6}Li(p,γ)^{7}Be cross section discovered a resonance-like structure around 195 keV [43]. The observed S-factor is roughly reproduced by an R-matrix fit assuming the existence of an excited state with E $\sim$ 5800 keV, J$^{\pi}$ = (1/2^{+}, 3/2^{+}) and $\Gamma_p \sim$ 50 keV.
In order to verify the existence of such state, a simultaneous measurement of the ^{6}Li(p,γ)^{7}Be and ^{6}Li(p,α)^{3}He reactions has been performed at the LUNA solid target beam line. Gamma rays were detected using a HPGe detector in close geometry at 55°, while α and ^{3}He particles were detected by a Si detector at 125°. Targets of three different compositions were used (lithium oxide, lithium tungstate and lithium chloride) in order to evaluate possible systematic effects due to target composition and thickness. The data taking was completed in September 2017 and the analysis is presently ongoing.

A rich program is still scheduled for the LUNA 400 kV accelerator. The reactions ^{13}C(α,n)^{16}O, ^{22}Ne(α,γ)^{26}Mg, relevant for s-process nucleosynthesis, are presently under investigation while (p,γ) reactions on ^{12}C and ^{13}C are scheduled for the future.

5. The LUNA-MV project

The future of LUNA involves the installation of a new 3.5-MV accelerator. This machine will be devoted to the study of nuclear reactions that take place at higher temperatures than those occurring during the hydrogen-burning processes studied so far at the 50-kV and 400-kV facilities.
The new accelerator will provide high-intensity hydrogen, helium and carbon beams [44]. Two different beam lines will be built, one devoted to solid target and the other to gas

target experiments. This will allow the study of key reactions of helium burning (namely the ^{12}C(α,γ)^{16}O reaction and (α,n) reactions on ^{12}C and ^{22}Ne) and to further investigate the ^{14}N(p,γ)^{15}O and the ^{3}He(^{4}He,γ)^{7}Be reactions, already measured at the LUNA 400-kV facility, over a wide energy range in order to reduce their experimental uncertainties.
The LUNA-MV building will be surrounded by a 80 cm thick concrete neutron shielding to preserve the low neutron background conditions of LNGS.
The construction of the infrastructure is in progress and the installation of the accelerator is scheduled for the next year.

References

[1] Rolfs C E and S R W 1988 *Cauldrons in the Cosmos* (The University of Chicago Press) ISBN 978-0-266-72457-3
[2] Boeltzig A *et al.* 2018 *Journal of Physics G: Nuclear and Particle Physics* **45** 025203
[3] Best A *et al.* 2016 *The European Physical Journal A* **52** 72 ISSN 1434-601X
[4] The LUNA Collaboration, Szücs T *et al.* 2010 *The European Physical Journal A* **44** 513–519 ISSN 1434-601X
[5] Robertson, Daniel, Couder, Manoel, Greife, Uwe, Strieder, Frank and Wiescher, Michael 2016 *EPJ Web of Conferences* **109** 09002 URL `https://doi.org/10.1051/epjconf/201610909002`
[6] Wang S and Li K 2018 *AIP Conference Proceedings* **1947** 020004
[7] Bemmerer D *et al.* 2019 *Proceedings of the 5th International Solar Neutrino Conference* ed Meyer M and Zuber K (Singapore: World Scientific) (*Preprint* 1810.08201)
[8] Caciolli A, Agostino L, Bemmerer D, Bonetti R, Broggini C, Confortola F, Corvisiero P, Costantini H, Elekes Z, Formicola A, Fülöp Z, Gervino G, Guglielmetti A, Gustavino C, Gyürky G, Imbriani G, Junker M, Laubenstein M, Lemut A, Limata B, Marta M, Mazzocchi C, Menegazzo R, Prati P, Roca V, Rolfs C, Rossi Alvarez C, Somorjai E, Straniero O, Strieder F, Terrasi F and Trautvetter H P 2009 *Eur. Phys. J. A* **39** 179–186
[9] Greife U *et al.* 1994 *Nucl. Instr. Meth. Phys. Res. A* **350** 327
[10] Formicola A *et al.* 2003 *Nucl. Instr. Meth. Phys. Res. A* **507** 609–616
[11] Formicola, Alba, Corvisiero, Pietro and Gervino, Gianpiero 2016 *Eur. Phys. J. A* **52** 73
[12] Gustavino, C, Anders, M, Bemmerer, D, Elekes, Z and Trezzi, D 2016 *Eur. Phys. J. A* **52** 74
[13] Boeltzig A *et al.* 2016 *The European Physical Journal A* **52** 75 ISSN 1434-601X
[14] Broggini C, Bemmerer D, Caciolli A and Trezzi D 2018 *Progress in Particle and Nuclear Physics* **98** 55 – 84 ISSN 0146-6410
[15] Cavanna F and Prati P 2018 *International Journal of Modern Physics A* **33** 1843010
[16] Scott D A *et al.* (LUNA Collaboration) 2012 *Phys. Rev. Lett.* **109**(20) 202501
[17] Berheide M, Rolfs C, Schröder U and Trautvetter H P 1992 *Zeitschrift für Physik A Hadrons and Nuclei* **343** 483–487 ISSN 0939-7922
[18] Niemeyer S 1996 *Diplomarbeit* Ruhr–Universität Bochum
[19] Blackmon J C *et al.* 1995 *Phys. Rev. Lett.* **74**(14) 2642–2645
[20] Angulo C *et al.* 1999 *Nucl. Phys. A* **656** 3–187
[21] Iliadis C, Longland R, Champagne A E, Coc A and Fitzgerald R 2010 *Nucl. Phys. A* **841** 31–250 (*Preprint* 1004.4517)
[22] Sergi M L *et al.* 2015 *Phys. Rev. C* **91**(6) 065803
[23] LUNA Collaboration, Bruno C G *et al.* 2015 *The European Physical Journal A* **51** 94 ISSN 1434-601X
[24] Bruno C G *et al.* 2016 *Phys. Rev. Lett.* **117** 142502 (*Preprint* 1610.00483)
[25] Straniero, O *et al.* 2017 *A&A* **598** A128
[26] Lugaro M *et al.* 2017 *Nature Astronomy* **1** 0027
[27] Carretta E *et al.* 2009 *Astron. Astrophys.* **505** 117–138 (*Preprint* 0909.2938)
[28] Izzard R G *et al.* 2007 *Astron. Astrophys.* **466** 641–648 (*Preprint* arXiv:astro-ph/0703078)
[29] Iliadis C, Champagne A, José J, Starrfield S and Tupper P 2002 *Astrophys. J. Suppl. Ser.* **142** 105–137
[30] Depalo R *et al.* 2015 *Phys. Rev. C* **92** 045807 (*Preprint* 1507.03893)
[31] Kelly K J *et al.* 2017 *Phys. Rev. C* **95**(1) 015806
[32] LUNA Collaboration, Cavanna F *et al.* 2014 *The European Physical Journal A* **50** 179 ISSN 1434-601X
[33] Cavanna F *et al.* (The LUNA Collaboration) 2015 *Phys. Rev. Lett.* **115**(25) 252501
[34] Depalo R *et al.* (LUNA Collaboration) 2016 *Phys. Rev. C* **94**(5) 055804
[35] Ferraro F *et al.* (LUNA Collaboration) 2018 *Phys. Rev. Lett.* **121**(17) 172701
[36] Hale S E *et al.* 2001 *Phys. Rev. C* **65** 015801
[37] andothers C B 2019 *Physics Letters B* **790** 237 – 242 ISSN 0370-2693

IOP Conf. Series: Journal of Physics: Conf. Series **1308** (2019) 012009 doi:10.1088/1742-6596/1308/1/012009

[38] Cooke R J, Pettini M and Steidel C C 2018 *The Astrophysical Journal* **855** 102 URL
[39] Planck Collaboration, Aghanim N *et al.* 2018 *arXiv e-prints* (*Preprint* 1807.06209)
[40] Adelberger E G *et al.* 2011 *Rev. Mod. Phys.* **83** 195–246 (*Preprint* 1004.2318)
[41] LUNA Collaboration, Ferraro, F *et al.* 2018 *Eur. Phys. J. A* **54** 44
[42] Cruz J *et al.* 2008 *Journal of Physics G: Nuclear and Particle Physics* **35** 014004
[43] He J *et al.* 2013 *Physics Letters B* **725** 287 – 291 ISSN 0370-2693
[44] Sen A *et al.* 2018 *Nucl. Instr. Meth. Phys. Res. B* ISSN 0168-583X

IOP Conf. Series: Journal of Physics: Conf. Series **1308** (2019) 012010 doi:10.1088/1742-6596/1308/1/012010

Dark particle interpretation of the neutron decay anomaly

Bartosz Fornal and Benjamin Grinstein

Department of Physics, UC San Diego, La Jolla, California 92093, USA

E-mail: bfornal@ucsd.edu, bgrinstein@ucsd.edu

Abstract. There is a long-standing discrepancy between the neutron lifetime measured in beam and bottle experiments. We propose to explain this anomaly by a dark decay channel for the neutron, involving a dark sector particle in the final state. If this particle is stable, it can be the dark matter. Its mass is close to the neutron mass, suggesting a connection between dark and baryonic matter. In the most interesting scenario a monochromatic photon with energy in the range 0.782 MeV – 1.664 MeV and branching fraction 1% is expected in the final state. We construct representative particle physics models consistent with all experimental constraints.

1. Introduction

In the previous talk, A. Holley presented the status of measurements of neutron lifetime. We will not repeat this here. The purpose of this talk is to contemplate the possibility that the discrepancy between neutron lifetime measurements in bottle and beam experiments is due to a new phenomenon, rather than a systematic error or a statistical fluctuation [1]. The measurements differ by about 1%; in terms of the decay widths, $\Delta\Gamma_n^{\rm exp} = \Gamma_n^{\rm bottle} - \Gamma_n^{\rm beam} \simeq 7.1 \times 10^{-30}$ GeV $\simeq (9.5 \times 10^{-3})\Gamma_n^{\rm bottle}$.

Basic principles demand that the particle lifetime does not depend on the measurement technique or decay mode. For a new physics explanation of the discrepancy we posit that one of the experiments is not reporting a lifetime. The bottle experiment fits to a decaying in time exponential, measuring the lifetime by definition. An important observation is that the beam experiment, by contrast, measures the instantaneous β-decay rate:

$$\tau_n^{\rm beam} = \frac{\tau_n}{{\rm Br}(n \to p + {\rm anything})} \,. \qquad (1)$$

This gives the decay lifetime only if the neutron β-decays 100% of the time. Therefore, the discrepancy could be explained if the neutron decays about 1% of the time into a protonless final state. A very direct test of this hypothesis could be performed, but has not been done, in both bottle and beam experiments. In a bottle experiment one could measure the number of decays that result in final state protons: the hypothesis asserts that these would account only for 99% of the neutrons lost. In a beam experiment one could fit the number of observed protons to an exponential decaying along the beam: here the fit to the exponential would result in a 1% smaller lifetime than the one determined from the instantaneous rate at one point along the beam.

The question immediately arises as to what is the new decay channel of the neutron, a protonless state with 1% branching fraction. It is incumbent upon us to present a scenario that is consistent with our current understanding of elementary particle interactions.

2. Model independent analysis

The final state of neutron decay must have unit fermion number and be overall electrically neutral. The simplest possibility for a protonless decay channel consists of a chargeless, "dark" fermion, χ, and a neutral boson. In principle the dark fermion could be a neutrino and the neutral boson could be a photon, but soon we will see that the sum of their masses must be within 1.572 MeV of the neutron mass, so at least one of the two must be a new elementary particle. We consider two possible final states: (i) a dark fermion, χ, and a photon, and (ii) a dark fermion, χ, and a new spinless elementary dark boson, ϕ.

In either case, the mass M_f of the final state f must be sufficiently large that proton stability is not vitiated: to avoid $p \to n^* e^+ \nu$ followed by $n^* \to f$, we must have $M_f > m_p - m_e$. A slightly stronger bound follows from stability of ^{9}Be against decay to $2\,^4\text{He} + f$, yielding [2]:

$$937.992 \text{ MeV} < M_f < 939.565 \text{ MeV} . \tag{2}$$

For the case $f = \chi + \gamma$ this immediately gives 937.992 MeV $< m_\chi <$ 939.565 MeV. For $m_\chi < m_p + m_e = 938.783$ MeV, β decay of χ is not allowed giving the interesting possibility that χ is stable (or very long lived) and could be a candidate for a dark matter component of the universe. Similarly, for the case $f = \chi + \phi$ both χ and ϕ are dark matter candidates if $|m_\chi - m_\phi| < m_p + m_e$.

2.1. *Effective theory for $\chi\gamma$ final state*

To describe the decay $n \to \chi\gamma$ in a quantitative way, we consider theories with a mass mixing term $\chi\, n$, and an induced interaction $\chi\, n\, \gamma$. An example of such a theory is given by the effective Lagrangian

$$\mathcal{L}_1^{\text{eff}} = \bar{n}\left(i\partial\!\!\!/ - m_n + \tfrac{g_n e}{2m_n}\sigma^{\mu\nu}F_{\mu\nu}\right)n + \bar{\chi}\left(i\partial\!\!\!/ - m_\chi\right)\chi + \varepsilon\left(\bar{n}\chi + \bar{\chi}n\right) , \tag{3}$$

where $g_n \simeq -3.826$ is the neutron g-factor and ε is the mixing parameter with dimension of mass. The term corresponding to $n \to \chi\gamma$ is obtained by transforming Eq. (3) to the mass eigenstate basis and, for $\varepsilon \ll m_n - m_\chi$, yields

$$\mathcal{L}^{\text{eff}}_{n\to\chi\gamma} = \frac{g_n e}{2m_n}\frac{\varepsilon}{(m_n - m_\chi)}\,\bar{\chi}\,\sigma^{\mu\nu}F_{\mu\nu}\,n . \tag{4}$$

Therefore, the neutron dark decay rate is

$$\Delta\Gamma_{n\to\chi\gamma} = \frac{g_n^2 e^2}{8\pi}\left(1 - \frac{m_\chi^2}{m_n^2}\right)^3 \frac{m_n\,\varepsilon^2}{(m_n - m_\chi)^2} \approx \Delta\Gamma_n^{\text{exp}}\left(\tfrac{1+x}{2}\right)^3\left(\tfrac{1-x}{1.8\times10^{-3}}\right)\left(\tfrac{\varepsilon\ [\text{GeV}]}{9.3\times10^{-14}}\right)^2,$$

where $x = m_\chi/m_n$. The rate is maximized when m_χ saturates the lower bound, $m_\chi =$ 937.992 MeV. Below we will give a microscopic particle physics realization of this case.

The testable prediction of this class of models is a monochromatic photon with an energy in the range $0 < E_\gamma < 1.572$ MeV and a branching fraction

$$\frac{\Delta\Gamma_{n\to\chi\gamma}}{\Gamma_n} \approx 1\% .$$

If the dark fermion χ is to be sufficiently light that it may be stable, and hence a DM candidate, then 0.782 MeV $< E_\gamma <$ 1.572 MeV. A null search for monochromatic photons rules out this

IOP Conf. Series: Journal of Physics: Conf. Series **1308** (2019) 012010 doi:10.1088/1742-6596/1308/1/012010

hypothesis in the range 0.782 to 1.664 MeV at the 97% confidence level [3]. A signature involving an e^+e^- pair with total energy $E_{e^+e^-} < 1.572$ MeV is also expected, but with a suppressed branching fraction of at most 1.1×10^{-6}. A null search for $n \to \chi e^+e^-$ sets a bound of $\sim 10^{-4}$ on the branching fraction with 99.9997% significance for 937.900 MeV $< m_\chi <$ 938.543 MeV [4].

2.2. Effective theory for $\chi\phi$ final state

Consider now a scenario in which the neutron decays to an intermediate off-shell dark fermion, $\tilde{\chi}$, which subsequently decays to a final state dark fermion and a scalar, χ and ϕ. The masses of the final state particles are constrained by Eq. (2), with $M_f = m_\chi + m_\phi$, and both χ and ϕ are stable if $|m_\chi - m_\phi| <$ 938.783 MeV. It is also possible to have an on-shell decay to $\tilde{\chi}\gamma$ if 937.992 MeV $< m_{\tilde{\chi}} <$ 939.565 MeV, but we are mostly interested in the case $m_{\tilde{\chi}} >$ 939.565 MeV for which the neutron decay is dark.

An example of such a theory is

$$\mathcal{L}_2^{\rm eff} = \bar{n}\left(i\partial\!\!\!/ - m_n + \tfrac{g_n e}{2m_n}\sigma^{\mu\nu}F_{\mu\nu}\right)n + \bar{\tilde{\chi}}\left(i\partial\!\!\!/ - m_{\tilde{\chi}}\right)\tilde{\chi} + \bar{\chi}\left(i\partial\!\!\!/ - m_\chi\right)\chi$$
$$+ \partial_\mu\phi^*\partial^\mu\phi - m_\phi^2|\phi|^2 + \varepsilon\left(\bar{n}\tilde{\chi} + \bar{\tilde{\chi}}n\right) + \left(\lambda_\phi\,\bar{\tilde{\chi}}\,\chi\,\phi + \text{h.c.}\right) . \quad (5)$$

Going to the mass eigenstate basis, the term corresponding to $n \to \chi\phi$ is

$$\mathcal{L}^{\rm eff}_{n\to\chi\phi} = \frac{\lambda_\phi\,\varepsilon}{m_n - m_{\tilde{\chi}}}\,\bar{\chi}\,n\,\phi^* , \quad (6)$$

which yields the neutron dark decay rate

$$\Delta\Gamma_{n\to\chi\phi} = \frac{|\lambda_\phi|^2}{16\pi}\sqrt{f(x,y)}\,\frac{m_n\,\varepsilon^2}{(m_n - m_{\tilde{\chi}})^2} , \quad (7)$$

where $f(x,y) = [(1-x)^2 - y^2]\,[(1+x)^2 - y^2]^3$ with $x = m_\chi/m_n$ and $y = m_\phi/m_n$. A microscopic, particle physics realization of this scenario is provided below.

For $m_{\tilde{\chi}} > m_n$ the missing energy signature has a branching fraction $\approx 1\%$. There will also be a very suppressed radiative process involving a photon in the final state with a branching fraction 3.5×10^{-10} or smaller.

As discussed earlier, in the case 937.992 MeV $< m_{\tilde{\chi}} < m_n$ both the visible and invisible neutron dark decay channels are present. The ratio of their branching fractions is

$$\frac{\Delta\Gamma_{n\to\tilde{\chi}\gamma}}{\Delta\Gamma_{n\to\chi\phi}} = \frac{2g_n^2 e^2}{|\lambda_\phi|^2}\frac{(1-\tilde{x}^2)^3}{\sqrt{f(x,y)}} , \quad (8)$$

where $\tilde{x} = m_{\tilde{\chi}}/m_n$, while their sum accounts for the neutron decay anomaly, i.e.

$$\frac{\Delta\Gamma_{n\to\tilde{\chi}\gamma} + \Delta\Gamma_{n\to\chi\phi}}{\Gamma_n} \approx 1\% . \quad (9)$$

The branching fraction for the process involving a photon in the final state ranges from about 0 to 1%, depending on the masses and couplings. A suppressed decay channel involving e^+e^- is also present.

3. Microscopic Models

While the effective theory analysis presented above is useful in estimating the dark decay parameters that a fundamental microscopic model must produce, the existence of a viable model with renormalizable interactions that satisfies all experimental constraints is not obvious. We will

IOP Conf. Series: Journal of Physics: Conf. Series **1308** (2019) 012010 doi:10.1088/1742-6596/1308/1/012010

present one example of a microscopic model for each of the scenarios above. Other possibilities may exist, but we are satisfied for now by presenting these working examples.

The minimal model for the $n \to \chi\gamma$ requires only two particles beyond the SM: a scalar $\Phi = (3,1)_{-1/3}$ (color triplet, weak singlet, hypercharge $-1/3$), and a Dirac fermion χ (SM singlet). The interaction Lagrangian of the model includes

$$\mathcal{L}_1 = \lambda_q\, \epsilon^{ijk}\, \overline{u^c_{L i}}\, d_{Rj} \Phi_k + \lambda_\chi \Phi^{*i} \bar{\chi}\, d_{Ri} + \text{h.c.}\,, \tag{10}$$

where u^c_L is the charge conjugate of u_R. Assigning baryon numbers $B_\chi = 1$, $B_\Phi = -2/3$ we see that proton decay is forbidden [5–7]. The rate for $n \to \chi\gamma$ is given by Eq. (5) with

$$\varepsilon = \frac{\beta\, \lambda_q \lambda_\chi}{M_\Phi^2} \tag{11}$$

and β defined by $\langle 0|\epsilon^{ijk}\left(\overline{u^c_{Li}} d_{Rj}\right) d^\rho_{Rk}|n\rangle = \beta \frac{1}{2}\left(1+\gamma_5\right)^\rho{}_\sigma\, u^\sigma$, where u is the neutron spinor, σ is the spinor index and the parenthesis denote spinor contraction. Lattice QCD calculations give $\beta = 0.0144(3)(21)$ GeV^3 [8], where the errors are statistical and systematic, respectively. Assuming $m_\chi = 937.992$ MeV to maximize the rate, in order to explain the anomaly the parameters must satisfy

$$\frac{|\lambda_q \lambda_\chi|}{M_\Phi^2} \approx 6.7 \times 10^{-6}\ \text{TeV}^{-2}\,. \tag{12}$$

In addition to the monochromatic photon with energy $E_\gamma < 1.572$ MeV and the e^+e^- signal, one may search directly also for Φ. It can be singly produced through $p\,p \to \Phi$ or pair produced via gluon fusion $g\,g \to \Phi\,\Phi$. This results in a dijet or four-jet signal from $\Phi \to d^c u^c$, as well as a monojet plus missing energy signal from $\Phi \to d\,\chi$. Given Eq. (12), Φ is not excluded by recent LHC analyses [9–14] provided $M_\Phi \gtrsim 1$ TeV.[1]

The parameter choice in Eq. (12) is excluded if χ is a Majorana particle, as in the model proposed in [15], by the neutron-antineutron oscillation and dinucleon decay constraints [16,17]. Neutron decays considered in [18] are too suppressed to account for the neutron decay anomaly.

A representative model for the case $n \to \chi\phi$ involves four new particles: the scalar $\Phi = (3,1)_{-1/3}$, two Dirac fermions $\tilde{\chi}$, χ, and a complex scalar ϕ, the last three being SM singlets. The interaction Lagrangian is an extension of the one in (10)

$$\mathcal{L}_2 = \mathcal{L}_1(\chi \to \tilde{\chi}) + (\lambda_\phi\, \bar{\tilde{\chi}}\, \chi\, \phi + \text{h.c.})\,. \tag{13}$$

We have imposed an additional $U(1)$ symmetry under which χ and ϕ have opposite charges. For $m_\chi > m_\phi$ the annihilation channel $\chi\,\bar{\chi} \to \phi\,\bar{\phi}$ via a t-channel $\tilde{\chi}$ exchange is open. The observed DM relic density, assuming $m_\chi = 937.992$ MeV and $m_\phi \approx 0$, is obtained for $\lambda_\phi \simeq 0.037$. Alternatively, the DM can be non-thermally produced.

The rate for $n \to \chi\,\phi$ is described by Eq. (7) with $\varepsilon = \beta\,\lambda_q\lambda_\chi/M_\Phi^2$. For $m_{\tilde{\chi}} = m_\chi$, the anomaly is explained with

$$\frac{|\lambda_q \lambda_\chi|}{M_\Phi^2}\frac{|\lambda_\phi|}{0.04} \approx 4.9 \times 10^{-7}\ \text{TeV}^{-2}. \tag{14}$$

For $\lambda_\phi \approx 0.04$ this is consistent with LHC searches, provided again that $M_\Phi \gtrsim 1$ TeV. Direct DM detection searches present no constraints. For similar reasons as before, χ and $\tilde{\chi}$ cannot be Majorana particles.

[1] A similar model with a scalar $\Phi = (3,1)_{2/3}$ and a Dirac fermion χ would also work. Since the scalar $(3,1)_{2/3}$ cannot couple to two first generation quarks, the rate in Eq. (12) would be suppressed by the strange quark content of the neutron and would require a larger value of $|\lambda_q\lambda_\chi|/M_\Phi^2$. Another viable option for Φ is the vector $(3,2)_{1/6}$.

IOP Conf. Series: Journal of Physics: Conf. Series **1308** (2019) 012010 doi:10.1088/1742-6596/1308/1/012010

As discussed above, in this model the branching fractions for the visible (including a photon) and invisible final states can be comparable, and their relative size is described by Eq. (8). A final state containing an e^+e^- pair is also possible. The same LHC signatures are expected as in model 1.

4. Further Developments

4.1. Unstable Nucleus Decay

While the lower bound in (2) with $M_f = m_\chi$ forbids the decay $(A, Z) \to (A-1, Z) + \chi$ for any stable nucleus, it is still permitted for some unstable nuclei. There are several unstable isotopes with a neutron binding energy $S(n) < 1.572$ MeV and a sufficiently long lifetime to probe the dark decay channel when the dark particle mass $m_\chi < m_n - S(n)$. One example presented in [1] is ^{11}Li, for which $S(n) = 0.396$ MeV. ^{11}Li β decays with a lifetime 8.75 ms. However, in the presence of a dark particle χ the decay chain $^{11}\mathrm{Li} \to {}^{10}\mathrm{Li} + \chi \to {}^{9}\mathrm{Li} + n + \chi$ becomes available. Ref. [2] proposes to search for $^{11}\mathrm{Be} \to^{10} \mathrm{Be} + \chi$. This has $S(n) = 501.6(3)$ keV and half-life 13.76(7) s. Since this is a halo neutron nucleus, the calculation of the decay rate $^{11}\mathrm{Be} \to^{10} \mathrm{Be} + \chi + \gamma$ can be well estimated from that of $n \to \chi + \gamma$ by accounting for the phase-space factor. This leads to a branching fraction of $10^{-6} - 10^{-4}$ for 938 MeV $\lesssim m_\chi \lesssim$ 939 MeV. However, the dark decay of ^{11}Be does not necessarily involve a photon in the final state. For example, if the model (3) is supplemented by a π-nucleon interaction, $\pi\,\bar{n}n$, or a two-body nucleon interaction, $\bar{n}n\,\bar{n}n$, then in terms of the mass eigenstates one obtains interactions with the dark fermion, $\pi\,\bar{\chi}n$ and $\bar{\chi}n\,\bar{n}n$ that can directly mediate $(A, Z) \to (A-1, Z) + \chi$. Ref. [19] estimates the decay width for $^{11}\mathrm{Be} \to^{10} \mathrm{Be} + \chi$ to be an order of magnitude larger than the measured total width of ^{11}Be for $m_\chi \lesssim$ 939 MeV, excluding this model. However, we note that the result of Ref. [19] exhibits unphysical behavior: the decay amplitude, expressed as a function of $x = m_n - m_\chi - S(n)$, has singular behavior as $x \to 0$. No independent verification of the result is available. Yet, the computation uses the mixing term $\bar{\chi}n$, rather than the induced interactions, $\pi\,\bar{\chi}n$ or $\bar{\chi}n\,\bar{n}n$, to mediate the transition, and it is likely this artifact is responsible for the unphysical behavior.

The calculated branching fraction for dark ^{11}Be decay in Ref. [2] applies just as well to the purely dark decay of the model in (5). The branching fraction for ^{11}Be decay into ^{10}Be has been measured, and the result is some 400 times larger than that expected from β-delayed proton emission [20]. This could be explained if a yet unobserved very low lying and very narrow ^{11}B resonance existed. The authors of Ref. [2] propose an experiment measuring protons in ^{11}Be decay. This may discover the putative resonance, or, in the absence of the expected proton signal, give supporting evidence for the dark decay hypothesis.

4.2. Neutron Stars

The impact of neutron dark decays on neutron stars was considered in Refs. [21–23]. The resulting production of dark particles changes the energy density and pressure inside a neutron star, modifying its equation of state. This in turn changes the predictions for the maximum allowed neutron star masses, since they are derived from integrating the Tolman-Oppenheimer-Volkoff equation that explicitly depends on the equation of state.

It was shown that the observed neutron star masses ($2M_\odot$ for the heaviest neutron stars discovered) are allowed if strong repulsive self-interactions are present in the dark sector of our models. Such interactions are easily introduced in the representative Models 1 and 2 discussed in Sec. 3 by simply adding a dark vector boson coupled strongly to the dark particle χ.

Interestingly, a strongly self-interacting dark sector lies along the lines of the self-interacting dark matter paradigm, which was introduced two decades ago [24] to solve the core-cusp and missing satellite problem of the ΛCDM model.

A repulsive interaction between neutrons and dark particles can also be effective in allowing observed, $\lesssim 2M_\odot$, neutron star masses [25]. For appropriate strength and range of the interaction there is an energy cost associated with creating dark particles and therefore pure neutron matter is preferred.

4.3. Models with a self-interacting dark sector

A model of this type was constructed in Ref. [26], where a neutron dark decay involving a dark fermion and a dark photon in the final state was considered, i.e., $n \to \chi A'$. The effective Lagrangian is

$$\mathcal{L}^{\rm eff} = \bar{n}\left(i\not{D} - m_n + \tfrac{g_n e}{2m_n}\sigma^{\mu\nu}F_{\mu\nu}\right)n + \bar{\chi}\left(i\not{D} - m_\chi\right)\chi + \varepsilon\left(\bar{n}\,\chi + \bar{\chi}\,n\right) - \tfrac{1}{4}F'_{\mu\nu}F'^{\mu\nu} - \tfrac{\delta}{2}F_{\mu\nu}F'^{\mu\nu} - \tfrac{1}{2}m_{A'}^2 A'_\mu A'^\mu \,, \quad (15)$$

where the covariant derivative $D_\mu = \partial_\mu - i\,g' A'_\mu$. It was shown that the strength of the dark photon coupling to the dark particle χ, governed by the parameter g' and resulting in repulsive interactions between the χ particles, can be chosen such that the neutron lifetime discrepancy is explained and, at the same time, all astrophysical bounds are satisfied, including constraints from neutron stars, galaxy clusters, cosmic microwave background, Big Bang nucleosynthesis and supernovae. If the dark particle χ in this model is stable, it can contribute to the dark matter in the universe, but cannot account for all of the dark matter.

Many of the astrophysical constraints are alleviated if one assumes non-thermal dark matter production. This was shown in Ref. [27], where a model for the neutron dark decay $n \to \chi\,\phi$ was constructed, based on our Model 2, but with a dark boson introduced to mediate large self-interactions of χ. The Lagrangian for the dark sector is

$$\mathcal{L}_D = g\,\bar{\chi}\,\not{Z}_D\,\chi + (\lambda_\phi\,\bar{\tilde{\chi}}\,\chi\,\phi + {\rm h.c.}) - i\,g\,Z_D^\mu\left(\phi^*\partial_\mu\,\phi - \phi\,\partial_\mu\phi^*\right) \,. \quad (16)$$

There exists a choice of parameters for which this model satisfies neutron star constraints, remains consistent with all other astrophysical bounds and χ makes up all of the dark matter in the universe. In addition, due to the self-interactions of χ, the model is shown to solve the small-scale structure problems of the ΛCDM model.

4.4. Hadron dark decays

The idea of dark decays can be applied also to other neutral hadrons. In Ref. [28] it was argued that the mesons K_L^0 and B^0 can decay to dark sector particles at measurable rates. An explicit model was constructed with a dark sector consisting of several families of dark fermions. An analogous mechanism that prevents neutron beta decays in neutron stars, i.e., Pauli blocking, also forbids neutron dark decays inside a neutron star in this model.

4.5. Baryogenesis

It has recently been shown that the model addressing the neutron lifetime puzzle based on the Lagrangian in Eq. (15) provides a successful framework for low-scale baryogenesis [29]. In addition, a model very similar to our Model 2, with couplings of $\tilde{\chi}$ to other quark flavors and a Majorana (instead of Dirac) fermion χ, has been proposed in the context of low-scale baryogenesis as well [30].

4.6. Related solutions

Taking into consideration only the experimental data for g_A from experiments performed after the year 2002, the bottle neutron lifetime is favored [31]. Based on this observation, explanations

IOP Conf. Series: Journal of Physics: Conf. Series **1308** (2019) 012010 doi:10.1088/1742-6596/1308/1/012010

of the neutron lifetime discrepancy have been put forward in which it is the bottle lifetime that is equivalent to the Standard Model prediction for τ_n. The difference in outcomes of the bottle and beam measurements is explained via neutron-mirror neutron oscillations resonantly enhanced in large magnetic fields thus affecting only beam measurements [32], or by invoking a sizable Fierz interference term canceling the dark decay contribution to the neutron decay rate [33].

Acknowledgments
This research was supported in part by the DOE Grant No. DE-SC0009919.

References

[1] Fornal B and Grinstein B 2018 *Phys. Rev. Lett.* **120** 191801 (*Preprint* 1801.01124)
[2] Pfützner M and Riisager K 2018 *Phys. Rev.* **C97** 042501 (*Preprint* 1803.01334)
[3] Tang Z *et al.* 2018 *Phys. Rev. Lett.* **121** 022505 (*Preprint* 1802.01595)
[4] Sun X *et al.* (UCNA) 2018 *Phys. Rev.* **C97** 052501 (*Preprint* 1803.10890)
[5] Arnold J M, Fornal B and Wise M B 2013 *Phys. Rev.* **D87** 075004 (*Preprint* 1212.4556)
[6] Doršner I, Fajfer S, Greljo A, Kamenik J F and Košnik N 2016 *Phys. Rept.* **641** 1–68 (*Preprint* 1603.04993)
[7] Assad N, Fornal B and Grinstein B 2018 *Phys. Lett.* **B777** 324–331 (*Preprint* 1708.06350)
[8] Aoki Y, Izubuchi T, Shintani E and Soni A 2017 *Phys. Rev.* **D96** 014506 (*Preprint* 1705.01338)
[9] Aaboud M *et al.* (ATLAS) 2017 *Phys. Rev.* **D96** 052004 (*Preprint* 1703.09127)
[10] Sirunyan A M *et al.* (CMS) 2017 *Phys. Lett.* **B769** 520–542 [Erratum: Phys. Lett.B772,882(2017)] (*Preprint* 1611.03568)
[11] Aaboud M *et al.* (ATLAS) 2018 *Eur. Phys. J.* **C78** 250 (*Preprint* 1710.07171)
[12] Khachatryan V *et al.* (CMS) 2015 *Phys. Lett.* **B747** 98–119 (*Preprint* 1412.7706)
[13] Aaboud M *et al.* (ATLAS) 2018 *JHEP* **01** 126 (*Preprint* 1711.03301)
[14] Sirunyan A M *et al.* (CMS) 2017 *JHEP* **07** 014 (*Preprint* 1703.01651)
[15] McKeen D and Nelson A E 2016 *Phys. Rev.* **D94** 076002 (*Preprint* 1512.05359)
[16] Abe K *et al.* (Super-Kamiokande) 2015 *Phys. Rev.* **D91** 072006 (*Preprint* 1109.4227)
[17] Gustafson J *et al.* (Super-Kamiokande) 2015 *Phys. Rev.* **D91** 072009 (*Preprint* 1504.01041)
[18] Davoudiasl H 2015 *Phys. Rev. Lett.* **114** 051802 (*Preprint* 1409.4823)
[19] Ejiri H and Vergados J D 2019 *J. Phys.* **G46** 025104 (*Preprint* 1805.04477)
[20] Riisager K *et al.* 2014 *Phys. Lett.* **B732** 305–308 (*Preprint* 1402.1645)
[21] McKeen D, Nelson A E, Reddy S and Zhou D 2018 *Phys. Rev. Lett.* **121** 061802 (*Preprint* 1802.08244)
[22] Baym G, Beck D H, Geltenbort P and Shelton J 2018 *Phys. Rev. Lett.* **121** 061801 (*Preprint* 1802.08282)
[23] Motta T F, Guichon P A M and Thomas A W 2018 *J. Phys.* **G45** 05LT01 (*Preprint* 1802.08427)
[24] Spergel D N and Steinhardt P J 2000 *Phys. Rev. Lett.* **84** 3760–3763 (*Preprint* astro-ph/9909386)
[25] Grinstein B, Kouvaris C and Nielsen N G 2018 (*Preprint* 1811.06546)
[26] Cline J M and Cornell J M 2018 *JHEP* **07** 081 (*Preprint* 1803.04961)
[27] Karananas G K and Kassiteridis A 2018 *JCAP* **1809** 036 (*Preprint* 1805.03656)
[28] Barducci D, Fabbrichesi M and Gabrielli E 2018 *Phys. Rev.* **D98** 035049 (*Preprint* 1806.05678)
[29] Bringmann T, Cline J M and Cornell J M 2019 *Phys. Rev.* **D99** 035024 (*Preprint* 1810.08215)
[30] Elor G, Escudero M and Nelson A 2019 *Phys. Rev.* **D99** 035031 (*Preprint* 1810.00880)
[31] Czarnecki A, Marciano W J and Sirlin A 2018 *Phys. Rev. Lett.* **120** 202002 (*Preprint* 1802.01804)
[32] Berezhiani Z 2018 (*Preprint* 1807.07906)
[33] Ivanov A N, Höllwieser R, Troitskaya N I, Wellenzohn M and Berdnikov Y A 2018 (*Preprint* 1806.10107)

IOP Conf. Series: Journal of Physics: Conf. Series **1308** (2019) 012011 doi:10.1088/1742-6596/1308/1/012011

The nuclei ^{12}C, ^{16}O and the role of the Pauli Exclusion Principle

P. O. Hess[1,2], M. Berriel-Aguayo[1] and L. J. Chávez-Nuñez[1]

[1]Instituto de Ciencias Nucleares, UNAM, Circuito Exterior, C.U., A.P. 70-543, 04510 México D.F., Mexico
and
[2]Frankfurt Institute for Advanced Studies FIAS and Institute for Theoretical Physics, J.W. Goethe - University Frankfurt, Germany

E-mail: [1]hess@nculeares.unam.mx

Abstract. We investigate the role of the Pauli Exclusion Principle (PEP) for light nuclei, at the examples of ^{12}C and ^{16}O. We show that ignoring the PEP does lead not only to a too dense spectrum at low energy but also to a wrong grouping into bands. Using a geometrical mapping, a triangular structure for ^{12}C and a tetrahedral structure in ^{16}O in the ground state is obtained by using the indistinguishably of the α-particles.

1. Introduction

The nuclei ^{12}C and ^{16}O play an important role in the investigation of α-cluster systems THSR,freer2017,funaki2010,schuck2013,schuck2017,funaki2018,review-clusters,schuck2018. No core shell model calculations have been performed [9, 10, 11] in order to understand, under several aspects, the structure of the Hoyle state. Also calculations within the *Antisymmetrized Molecular Dynamics* model (AMD)l [12, 13, 6] and within the *Orthogonality Condition Model* (OCM) [35] have been performed, using a microscopic Hamiltonian. These microscopic models lead to an understanding of the Hoyle state [15], important in the ^{12}C production in stars.

In [16, 17, 18] the *Algebraic Cluster model* (ACM) [19] has been applied to ^{12}C and ^{16}O, assuming in the first one a triangular structure and in the second one a tetrahedral structure. The ACM is purely algebraic and does not take into account the *Pauli Exclusion Principle* (PEP). The bands are obtained by rotating the classical geometrical structure and bands are differentiated according to their vibrational structure.

In [20] the ^{12}C nucleus was investigated within the *Semimicroscopic Algebraic Cluster Model* (SACM) [21, 22], which is also an algebraic model, but whose model space takes into account the PEP. The similarity to the ACM makes this model ideal to compare to the ACM. The main conclusions in [20] are: i) The association of bands

IOP Conf. Series: Journal of Physics: Conf. Series **1308** (2019) 012011 doi:10.1088/1742-6596/1308/1/012011

Role of the Pauli principle

in the ACM is not consistent, the projection to Pauli allowed states leads to a grouping of bands according to the $SU(3)$ shell model *irreducible representations* (irrep). ii) Not including the PEP leads to too many states at low energy. Doublets predicted in the ACM are not there any more. iii) The additional (second) 5^- state at low energy is not supported by the shell model. iv) the observation of the 5^- state in [16] does not prove a triangular structure, it is just part of a band which is obtained in all models. v) Applying a geometrical mapping, for the ground state band it is trivial that ^{12}C has a triangular and ^{16}O a tetrahedral structure. The reason is that the α-particles are indistinguishable and, thus, the distance between any two α particles must be the same. For higher bands different geometrical configurations may mix and the situation is not very clear.

In this contribution we will report in more detail on these findings and include some new results on ^{16}O. The complete results will be published elsewhere.

In Section 2 the SACM is briefly resumed and the Hamiltonian used for ^{12}C is presented. Results and the discussion will also be provided. Finally, in Section 3 conclusions will be drawn.

2. The model, Hamiltonian and Quadrupole operator

Within the SACM [21, 22] the clusters are treated within the shell model and are in their ground state. The whole nucleus is also treated within the shell model and is a sum of the two clusters with the addition of the relative motions between the clusters. Each relative motion is quantized by the relative oscillation creation and annihilation operators $\boldsymbol{\pi}^\dagger_{\lambda_k,m}$ and $\boldsymbol{\pi}_{\lambda_k,m}$, respectively, where λ_k refers to the k'th Jacobi coordinate. The generators of the relative motion in the k'th Jacobi coordinate are given by $\boldsymbol{C}^{m'}_m(k) = \boldsymbol{\pi}^\dagger_m(k)\boldsymbol{\pi}^{m'}(k)$ and the ones related to the sum over all relative vibrations are $\boldsymbol{C}^{m'}_m = \sum_k \boldsymbol{\pi}^\dagger_m(k)\boldsymbol{\pi}^{m'}(k) = \boldsymbol{\pi}^\dagger_m\boldsymbol{\pi}^{m'}$. Auxiliary scalar bosons $\boldsymbol{\sigma}^\dagger$ and $\boldsymbol{\sigma}$ are introduced in order to implement a cutoff value N, with $N = n_\pi + n_\sigma$, where n_π and n_σ are the number operators of the total number of π oscillations and of the number of σ-bosons. The σ-bosons have no physical meaning, save for the cut-off. The number of relative motion quanta in each Jacobi coordinate is limited from below by the *Wildermuth condition* [23], which is a minimal condition for satisfying the PEP. Multiplying the $SU(3)$ irreps of the clusters with the relative motion leads to a sum of total $SU(3)$ irreps which still in part do not satisfy the PEP. These are excluded by constructing the overlap of the list of irreps obtained by the list of the shell model $SU(3)$ irreps. This is a quite brief summary and for more details, please consult the original publications on the SACM [21, 22].

The model Hamiltonian proposed for ^{12}C is

IOP Conf. Series: Journal of Physics: Conf. Series **1308** (2019) 012011 doi:10.1088/1742-6596/1308/1/012011

Role of the Pauli principle

$$
\begin{aligned}
\boldsymbol{H} = & \hbar\omega \boldsymbol{n}_\pi - \chi \boldsymbol{\mathcal{C}}_2(\lambda,\mu) + t_2(\boldsymbol{\mathcal{C}}_2(\lambda,\mu))^2 + t\boldsymbol{\mathcal{C}}_3(\lambda,\mu) \\
& + \left(a + a_L(-1)^L + a_{Lnp}\Delta\boldsymbol{n}_\pi\right)\boldsymbol{L}^2 + b\boldsymbol{K}^2 \\
& + b_1\left[\left(\boldsymbol{\sigma}^\dagger\right)^2 - \left(\boldsymbol{\pi}^\dagger\cdot\boldsymbol{\pi}^\dagger\right)\right]\cdot[h.c.] \quad .
\end{aligned} \tag{1}
$$

The first term is just the harmonic oscillator field and the $\hbar\omega$ is fixed via $45\times A^{-1/3}-25\times A^{-2/3}$ [24], where its value is 14.89 for ^{12}C. The second term is related to the algebraic quadrupole-quadrupole interaction [25], with $\boldsymbol{\mathcal{C}}_2(\lambda,\mu)$ being the second order Casimir operator. The third and fourth terms allow for corrections in the relative ordering of $SU(3)$ irreps, where $\boldsymbol{\mathcal{C}}_3$ is the third order Casimir operator, which distinguishes between (λ,μ) and (μ,λ). The fifth term allows to describe changes in the moment of inertia for states with higher shell excitations and when the spin parity changes. The last term in the second line lifts the degeneracy in angular momentum for states within the same $SU(3)$ irrep. Up to here, the Hamiltonian is within the $SU(3)$ limit and permits analytic results, substituting the operators by their corresponding eigenvalues. In the last line, the term mixes $SU(3)$ irreps, it is a generator of a $O(4)$ group. The pure $SU(3)$ part has 7 free parameters, the same as in [16] (including N). For allowing the mixing a further parameter (b_1) is added, i.e., in the final calculations there are 8 parameters.

As the quadrupole transition operator we use the one given in [26], which is a symplectic generator, including connections to multiple $2\hbar\omega$ shell excitations. In an algebraic model it is customary to use the algebraic part of this operator, which does not connect shells, valid when inter-shell excitations are not considered. The physical quadrupole operator [26, 27] is given by

$$
\begin{aligned}
\boldsymbol{Q}^{phys}_{2m} &= Q^a_{2m} + \frac{\sqrt{6}}{2}\left(\boldsymbol{B}^\dagger_{2m} + \boldsymbol{B}_{2m}\right) \\
\boldsymbol{B}^\dagger_{2m} &= \left(\boldsymbol{\pi}^\dagger\cdot\boldsymbol{\pi}^\dagger\right) \;,\; \boldsymbol{B}_{2m} \;=\; \left(\boldsymbol{\pi}\cdot\boldsymbol{\pi}\right) \quad .
\end{aligned} \tag{2}
$$

The $\boldsymbol{B}^\dagger_{2m}$ operator transforms as a (2,0) $SU(3)$ irrep, while $\boldsymbol{B}_{2m}$ as its conjugate.

The states for the 3-α cluster nucleus ^{12}C and the 4-α cluster nucleus ^{16}O were constructed, using the work done in [29, 30]. The overlap of the symmetrized cluster states with the shell model is determined using spin-isospin zero supermultiplets [31] to which the total state is coupled. The microscopic model space for ^{12}C is given in [20] and for ^{16}O it will be published elsewhere.

In Fig. 1 several fits, each referring to the adjustment to the $B(E2;0_2^+\to 2_1^+)$-value, are depicted. The general agreement is quite good and comparable to the one given in [16]. However, the ordering into bands is completely different. The $SU(3)$ fit indicates that the bands are ordered according to definite $SU(3)$ irreps. In each irrep, the states have a definite deformation, as deduced in [32, 33]. Thus, intrinsic states with different deformation are associated into different bands!

IOP Conf. Series: Journal of Physics: Conf. Series **1308** (2019) 012011 doi:10.1088/1742-6596/1308/1/012011

Role of the Pauli principle

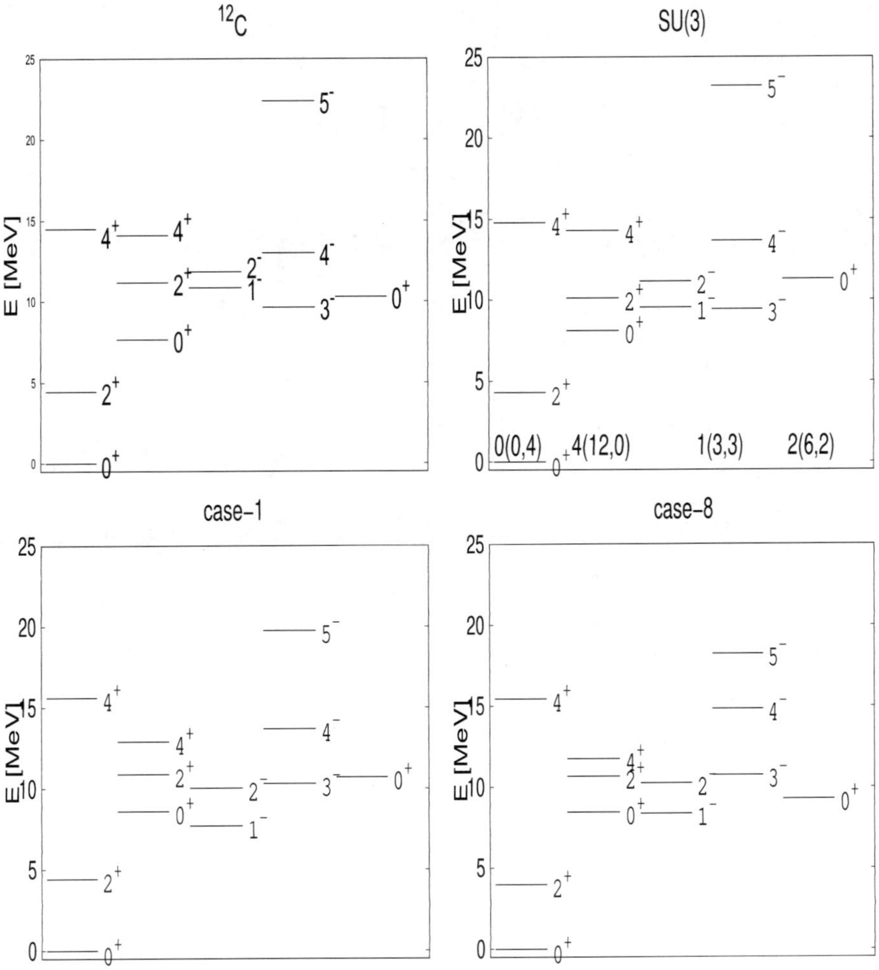

Figure 1. Spectrum of ^{12}C. In the upper row, the left figure is the experimental spectrum and the right figure depicts the result for pure $SU(3)$. In the second row, the left figure depicts the result when the $B(E2; 0_2^+ \rightarrow 2_1^+)$ transition value is adjusted to 1 WU and the right figure when this value is adjusted to the experimental one, namely 8 WU, which is the value from experiment [28]. The first excited 0^+ state corresponds to the hoyle state [15]. The experimental values are taken from [28]. In the $SU(3)$ plot, at the lower part the numbers $\Delta n_\pi(\lambda, \mu)$ are listed, where Δn_π refers to the shell exciations and (λ, μ) to the corresponding $SU(3)$ irrep.

In Fig. 2 we compare the case where the $B(E2; 2_1^+ \rightarrow 0_1^+)$ is adjusted to 1 WU (left panel) to the results of the ACM [16]. As seen, the spin-doublets are dissolved and the density of states is lower.

In Table 1 the $SU(3)$-content of the first two 0^+ bands are listed, where the values were rounded off. The association of states into the same band requires that the content is the same, however, in reality mixing effect reproduce only approximate agreements. As can be seen, the 2_1^+ state has a very similar content as the 0_1^+ state, thus, they belong to the same band. It is similar for the second 0^+ band. This cannot be concluded from

IOP Conf. Series: Journal of Physics: Conf. Series **1308** (2019) 012011 doi:10.1088/1742-6596/1308/1/012011

Role of the Pauli principle

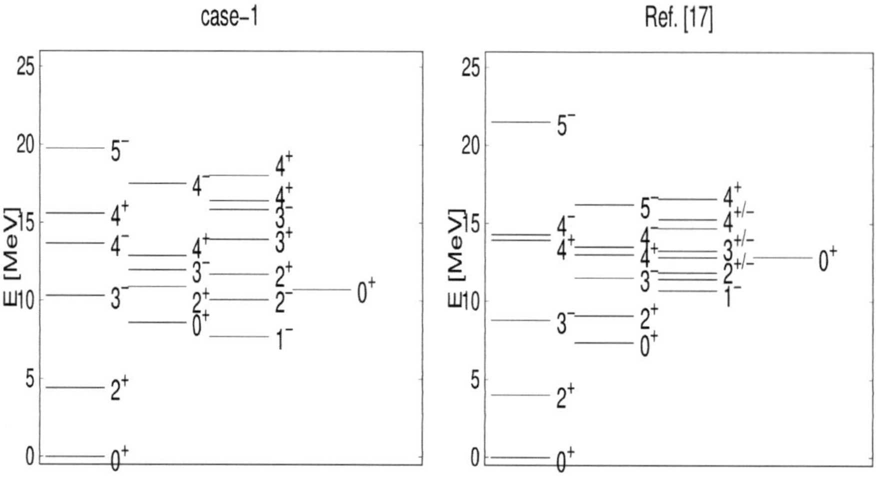

Figure 2. Comparison Spectrum of ^{12}C, as calculated within the SACm (left panel) to the AMC (right panel).

L_i^π	0_1^+	2_1^+	4_2^+	0_2^+	2_2^+	4_1^+
$n_\pi = 4$:(0,4)	82	82	48	8	3	31
$n_\pi = 6$:(2,4)	16	16	4	22	41	16
$n_\pi = 6$:(4,3)	0	0	0	0	0	0
$n_\pi = 8$:(4,4)	1	2	0	22	28	29
$n_\pi = 8$:(6,2)	0	0	0	5	6	0
$n_\pi = 8$:(6,3)	0	0	0	0	0	0
$n_\pi = 8$:(8,2)	0	0	0	2	6	0
$n_\pi = 8$:(12,0)	0	0	0	33	6	0
$n_\pi = 10$:(6,4)	0	0	39	7	9	19
$n_\pi = 10$:(10,2)	0	0	0	0	0	0

Table 1. $SU(3)$ content of some low lying states with positive parity, given as in percent, for the case of $B(E2; 0_2^+ \to 2_1^+) = 8\ WU$. The numbers are only approximate and not all irreps are shown.

$B(EL; J_i^\pi \to J_f^\pi)$ [WU]	EXP.	$SU(3)$	case-1	case-8
$B(E2; 2_1^+ \to 0_1^+)$	4.65	4.65	5.18	3.41
$B(E2; 0_2^+ \to 2_1^+)$	8.	0.0	0.97	8.33
$B(E3; 3_1^- \to 0_1^+)$	12.	6.32	5.27	24.28

Table 2. List of $B(EL)$-transition values, measured and obtained in three different model calculations: In the first column information is listed on the type of the electromagnetic transition, the second column lists the corresponding experimental value, the third column assumes exact $SU(3)$ symmetry and in the last two columns the experimental value of $B(E2; 2_2^+ \to 0_1^+)$ is adjusted to different values (case-1: 1WU, case-8: 8WU).

IOP Conf. Series: Journal of Physics: Conf. Series **1308** (2019) 012011 doi:10.1088/1742-6596/1308/1/012011

Role of the Pauli principle

$B(EL; J_i^\pi \to J_f^\pi)$ [WU]	EXP.	case-1	Ref. [16]
$B(E2; 2_1^+ \to 0_1^+)$	4.65 ± 0.25	5.18	5.15
$B(E2; 0_2^+ \to 2_1^+)$	8.03 ±1.10	0.97	0.80
$B(E3; 3_1^- \to 0_1^+)$	12.65 ± 1.99	12.23	5.15

Table 3. List of $B(EL)$-transition values, measured (first column), the one obtained within the SACM (second column) and within the ACM (third column). The numbers are given in WU.

the 4_1^+ and 4_2^+ states, which are already well mixed states, due to the large deformation of the ^{12}C nucleus. Here, the association into the bands is done in the eye of the beholder.

In Table 2 some experimental measured $B(E2)$ values and the $B(E3; 3_1^- \to 0_1^+)$ value are listed and compared to the $SU(3)$ calculation and two cases of theoretical fits, one where the side-band $B(E2; 0_2^+ \to 2_1^+)$ transition value is adjusted to 1 WU (as in [19]), called case-1, and another fit where this side-band transition value is adjusted to 8 WU, called case-8. The $B(E2)$-values were multiplied by a common scale factor $a \approx \frac{1}{2}$, which is proportional to the scale factor β^4 in [19]. The scale factor for the $B(E3)$ value is $a^{\frac{3}{2}}$, using the proportionality to β^6 [19]. The values for the $B(3)$ transition, listed in Table 2, were not multiplies by this factor. In the last column of Table 3 the corresponding $B(EL)$-values of the AMC, taken from [19], are listed and compared to the SACM and experimental values. In general, both theories are of the same quality. In the SACM the $B(E3; 3_1^- \to 0_1^+)$ is even better reproduced, it shows that the argument of a large $B(E3)$-value does not confirm the ACM. The main difference obviously shows up in the spectrum, suggesting that an effort has to be taken to measure if possible all states at low energy.

Why this difference to the ACM? In the ACM the α-particles are treated as independent classical objects and no antisymmetrization is performed. In ^{12}C they are located on the edges of a triangle and in ^{16}O on the edges of a tetrahedron. Upon this *classical* picture it is direct to create rotational bands, leading to the irreps of the corresponding discrete group. The problem with this is that the states are not antisymmetrized. By implementing the PEP these states are *projected* to Pauli allowed states, which destroys completely the ordering into irreps of the discrete groups. The final ordering is according to $SU(3)$ irreps in a pure $SU(3)$ model, and approximately when mixing is taken into account. The projection has nothing to do with a dynamical symmetry breaking within an algebraic model, because the states obtained in such a manner have still contributions which are not Pauli allowed.

An important feature, proven in [34, 35], is that the antisymmetrization between two clusters (PEP) produces an effective repulsive potential, i.e., that the fermions (nucleons) try to avoid to be at the same place This was later shown for two clusters in a different context [36], where after a geometrical mapping, the two clusters have a finite distance, a consequence that the Wildermuth condition [23], which requires a minimal number of oscillation quanta. Within the ACM, one can in principle simulate

IOP Conf. Series: Journal of Physics: Conf. Series **1308** (2019) 012011 doi:10.1088/1742-6596/1308/1/012011

the repulsive potential, however, adding high order terms in the generators. Because this is not done, the Pauli exclusion effect is not taken into account.

This can also be illustrated within a geometrical mapping, where a trial state $\mid \alpha\rangle$ is defined, with the the recommendation to use a coherent state. The expectation value of the Hamiltonian with respect to this trial state defines a classical potential. The description works fine for the ground state band. Using a coherent state, which describes α-clusters [36, 20] in the ground state, the ^{12}C has to be a triangle and ^{16}O has to be a tetrahedron. This is due to the indistinguishably of the α-particles: The distance between any two of those has to be the same. While for the ground state band in ^{12}C a triangular structure and in ^{16}O a tetrahedral structure is obtained, for the excited bands this is not necessarily the case due to the mixing of several geometrical configurations. Also, in the case of ^{12}C the AMC model shows a triangular structure in the density profile in one calculation [2], while the density profile is much more compact in another AMC calculation [37] and the nucleus looks like an oblate deformed nucleus.

As mentioned above, the projection to Pauli allowed states destroys this classical picture, the nuclei in consideration not only have an overlap to an α-cluster state but consist principally of nucleons which are in an antisymmetric state.

The conclusions drawn for ^{12}C are the same for ^{16}O. The difference is that while ^{12}C has a strong deformation, ^{16}O is spherical and is a showcase for the $SU(3)$ shell model. All states can be described within the SACM as nearly pure $SU(3)$ states. The ACM still predicts too many states at low energy (many doublets) which are not supported by the shell model. Details will be published elsewhere.

3. Conclusions

The SACM, which satisfies the PEP, was compared to the ACM, which does not consider the PEP. It was shown that in both systems, ^{12}C and ^{16}O, the PEP is still important due to the large overlap of the α-particles. Projecting the ACM states to Pauli allowed states destroys the association into bands as done in the ACM. The ACM obtains too many states/doublets at low energy and some which are not there. The concepts of atomic molecular molecules cannot be applied to nuclear molecules directly, however, one has to supplement it by the PEP.

In conclusion, the PEP cannot be neglected!

The ACM can be improved when the Pauli exclusion principle is taken into account via a repulsive potential between the α-particles, as discussed in the text. However, this requires high order interaction terms, which the ACM in its present form does not consider.

Acknowledgment

The authors acknowledge financial help from DGAPA-PAPIIT (IN100418) and CONACyT (Project no. 251817).

IOP Conf. Series: Journal of Physics: Conf. Series **1308** (2019) 012011 doi:10.1088/1742-6596/1308/1/012011

Role of the Pauli principle

[1] A. Tohsaki, H. Horiuchi, P. Schuck and G. Röpke, Phys. Rev. Lett. **87** (2001), 192501.
[2] M. Freer, Rep. Prog. Phys. **70** (2007) 2149.
[3] Y. Funaki, M Girod, H Horiuchi, G Röpke, P Schuck, A Tohsaki and T Yamada, J. Phys. G: Nucl. Part. Phys. **37** (2010) 064012.
[4] P Schuck, J. Phys.: Conf. Ser. **436** (2013), 012065
[5] P. Schuck, Y. Funaki, H. Horiuchi, G. Röpke, A. Tohsaki and T. Yamada, Phys. Srcip.**91** (2016), 123001
[6] Y. Funaki, Phys. Rev. C **97** (2018), 021304(R).
[7] M. Freer, H. Horiuchi, Y Kanada-En'yo, D. Lee and U.-G. Meissner, Rev. Mod. Phys. **90** (2018), 035004.
[8] P. Schuck, AIP Conference Proceedings **2038** (2018), 020002
[9] A.C. Dreyfuss, K.D. Launey, T. Dytrych, J.P. Draayer, C. Bahri, Phys. Lett. B **727** (2013), 511.
[10] T. Dytrych, *Evidence for Symplectic Symmetry in ab initio no-core Shell Model Results*, (PhD Thesis, Lousiana State Universiuty, 2008).
[11] K. D. Launey, T. Dytrych and J. P. Draayer, Progr. in Part. and Nucl. Phys. **89** (2016), 101.
[12] Y. Kanada-En'yo, H. Horiuchi, and A. Ono, Phys. Rev. C 52, 628 (1995)
[13] Y. Kanada-En'yo and H. Horiuchi, ibid. 52, 647 (1995).
[14] S. Saito, Prog. Theor. Phys. 41 (1969), 705.
[15] F. Hoyle, Ap. J. Suppl. **1** (1954), 121.
[16] D. J. Marín-Lámbarri, R. Bijker, M. Freer, et al., Phys. Rev. Lett. **113** (2014), 012502.
[17] R. Bijker and F. Iachello, Phys. Rev, Lett. **112** (2014), 152501.
[18] R. Bijker and F. Iachello, Nucl. Phys. A **957** (2017), 154.
[19] R. Bijker and F. Iachello, Ann. Phys. (N.Y.) **298** (2002), 334.
[20] P. O. Hess, Eur. Phys. J. A **54** (2018), 32.
[21] J. Cseh, Phys. Lett. B **281** (1992), 173.
[22] J. Cseh and G. Lévai, Ann. Phys. (N.Y.) **230** (1994), 165.
[23] K. Wildermuth and Y. C. Tang, *A Unified Theory of the Nucleus*, (Academic Press, New York, 1977).
[24] J. Blomqvist and A. Molinari, Nucl. Phys. A **106** (1968), 545.
[25] J. P. Elliott, Proc. R. Soc. London A **245**, (1958) 128; **245** (1958), 562.
[26] O. Castaños and J. P. Draayer, Nucl. Phys. A **491** (1989), 349.
[27] O. Castaños, P. O. Hess, P. Rocheford and J. P. Draayer, Nucl. Phys. A **524** (1991), 469.
[28] www.nndc.bnl.gov/ensdf
[29] M. Moshinsky and Y. Smirnoc. *The Harmonic Oscillator inModern Physics*, (Harwood Academic Publishers, Australia, 1996).
[30] V. C. Aguilera-Navarro, M. Moshinsky and P. Kramer, Ann. Phys. **54** (1969), 379.
[31] E. Wigner, in *Group Theoretical Concepts and Methods in Elementary Particle Physics*, ed. by. F. Gürsey., (Gordon and Breach, New York, 1964).
[32] D J Rowe, Rep. Prog. Phys. **48** (1985), 1419.
[33] O. Castaños, J. P. Draayer and Y. Leschber, Z. f. Phys. A**329** (1988), 33.
[34] S. Saito, Prog. Theor. Phys. **40** (1968), 893.
[35] S. Saito, Prog. Theor. Phys. **41** (1969), 705.
[36] P. O. Hess, G. Lévai and J. Cseh, Phys. Rev. C **54** (1996), 2345.
[37] Y. Kanada-En'yo, Phys. Rev. C **96** (2017), 034306.

IOP Conf. Series: Journal of Physics: Conf. Series **1308** (2019) 012012 doi:10.1088/1742-6596/1308/1/012012

THM applied to the investigation of explosive astrophysical scenarios.

M. La Cognata[1], S. Cherubini[1,2], M. Gulino[1,3], L. Lamia[1,2], R.G. Pizzone[1], S. Romano[1,2], C. Spitaleri[1], A. Tumino[1,3]

[1]Istituto Nazionale di Fisica Nucleare, Laboratori Nazionali del Sud, Via S. Sofia 62, Catania, 95123 Italy

[2]Dipartimento di Fisica e Astronomia "Ettore Majorana", Universitá di Catania, Via S. Sofia 64, Catania, 95123 Italy

[3]Facoltá di Ingegneria e Architettura, Universitá degli Studi di Enna "Kore", Cittadella Universitaria, Enna, 94100 Italy

E-mail: lacognata@lns.infn.it

Abstract. The Trojan Horse Method (THM) makes use of quasi-free reactions to deduce the cross section of nuclear reactions relevant for astrophysics at the energies of interest. Thanks to the suppression of the Coulomb barrier, the THM cross section does not exponentially vanishes at astrophysical energies. Here we will briefly summarise the fundamentals of the method, then we will discuss two applications of the method to reactions that have a pivotal role in the latest stages of stellar evolution, leading to explosive scenarios. In particular, we will focus on the indirect investigation of the $^{18}F(p,\alpha)^{15}O$ reaction, which is the most important ^{18}F destruction channel in novae, and the $^{12}C + ^{12}C$ reaction, which plays a critical role in astrophysics to understand stellar burning scenarios in carbon-rich environments, including supernovae.

1. The need of indirect techniques in nuclear astrophysics

Nuclear reaction cross sections $\sigma(E)$ have a pivotal role in modelling astrophysical phenomena since nuclear processes are responsible of energy production and synthesis of chemical elements in a variety of sites, ranging from the early universe to star's latest evolutionary stages. However, even at the highest astrophysical temperatures, reaching about 10^9 K, energies of interest are well below the Coulomb barriers of the fusing nuclei, in the case of reactions involving charged particles. Therefore, the Coulomb barrier strongly suppresses fusion cross sections, leading to vanishingly small signal-to-noise ratios. The building of underground laboratories has allowed, in few cases (see, for instance [1] for a review), to measure cross sections of astrophysical importance inside the energy window of astrophysical interest (the so-called Gamow window [2]). However, it turned out that at astrophysical energies, the presence of electrons cannot be neglected as the typical distances of closest approach are comparable to the atomic radii. The electron clouds surroundings projectile and target particles, usually in the form of partially stripped ions and atoms or molecules, respectively, screen the nuclear charges, determining an increase of the cross sections measured in the laboratory with respect to the case of bare nuclei. Such increase is described by an enhancement factor $f(E) = \exp\left(\pi\eta\frac{U_e}{E}\right)$, where U_e is the electron screening potential [3], which in the adiabatic limit (setting the upper limit for U_e), is calculated from the difference of the atomic binding energies in the fused system and in the separated atoms

[4]. However, many works (see, for instance, [5, 6, 7]) show that experimentally determined U_e values often exceed the theoretical upper limit, making it clear that our present understanding of the electron screening effect is far from being exhaustive.

Owing to the rapid change of cross sections with decreasing energies for charged-particle induced reactions, reaching values as small as picobarns, and to the uncertainties in magnitude due to the electron shielding, extrapolation, sometimes supported by theoretical calculations, is often the only possibility to have experimental estimates of nuclear cross sections at astrophysical energies. To this purpose, the astrophysical factor is usually adopted [3], $S(E) = \sigma(E)\, E\, \exp(2\pi\eta)$, where $\eta = \mathrm{Z_1 Z_2} e^2/\hbar v$ is the Sommerfeld parameter, $\mathrm{Z_1}$ and $\mathrm{Z_2}$ the atomic numbers of the interacting nuclei and v their relative velocity. Thanks to the introduction of the exponential factor, the main energy dependence of the cross section is compensated for and, as a consequence, $S(E)$ is a smooth function of the energy in the case of non resonant reactions.

However, resonant reactions play an important role in astrophysics as they might significantly increase the astrophysical factor, even by orders of magnitude, and alter nucleosynthesis and energy production, if resonances occur at energies of astrophysical interest. Therefore, unexpected resonances or inaccurate estimates of their strengths might introduce large systematic errors, making extrapolation very risky. A special role is played by broad sub threshold resonances. Even if resonance energies are lower than the entrance channel separation energy, resonance tails can be populated, causing an exponential increase of the astrophysical factor for center-of-mass energies approaching zero. Also in this case extrapolation is very challenging, owing to the unknown resonance strengths as well as to the interplay with the electron screening effect, also determining a steep rise of the astrophysical factor at energies close to the threshold. An important example is given by the measurement of the $^{13}\mathrm{C}(\alpha,n)^{16}\mathrm{O}$ S-factor (see, for instance, [8, 9, 10]), which is dominated by the tail of the $1/2^+$ $^{17}\mathrm{O}$ state at 6.356 MeV at astrophysical energies (140 − 230 keV).

It is also worth noting that the investigation of near threshold regions has important implications for nuclear structure. Indeed, resonances are caused by states in the intermediate compound nucleus formed in the reaction and their study makes it possible to deduce resonance energies, spin-parities and partial widths. Besides being difficult to investigate, the near threshold region is important as cluster structures are especially likely to appear right at these energies. Therefore, the introduction of alternative approaches such as indirect methods, in the exploration of nuclear reactions at astrophysical energies, has led to invaluable contribution in the understanding of nuclear astrophysics and nuclear structure. In what follows, we will focus on the Trojan horse method, established as a tool to investigate low-energy nuclear reactions by C. Spitaleri [11].

In recent years, the THM has been applied to reactions involved in quiescent nuclear burning, often involving low-energy proton-induced reactions (see, for instance, the recent review [12]). However, also in the case of explosive scenarios, when typical temperatures can be as large as 10^9 K and radioactive nuclei play an important role, THM represents a valuable technique to access the energies of astrophysical interest. Even if cross sections are generally larger than those characterising hydrostatic burning, yet the presence of high Z nuclei and/or radioactive species still makes direct measurement very challenging because of the small cross sections and of the low beam intensities, respectively.

2. Basic features of the resonant THM

The aim of the THM is to deduce the astrophysical factor of the $A(x,b)B$ reaction by studying a $A(a,bB)s$ process performed at energies of a few AMeV, namely, larger than the $A-a$ Coulomb barrier. Therefore, it is possible to bypass Coulomb and centrifugal barriers and electron screening effects (see [5] and references therein about the electron screening problem).

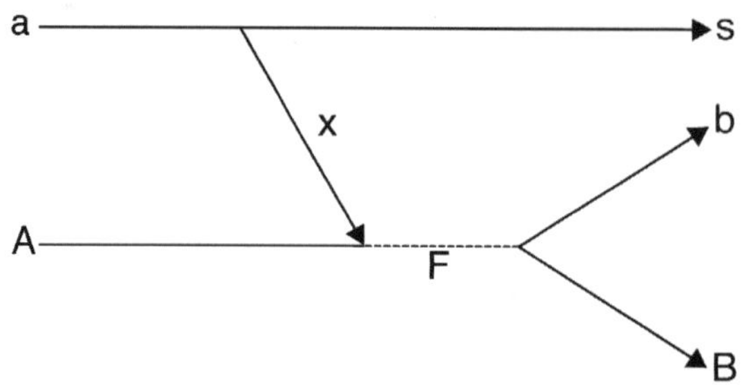

Figure 1. Sketch of the QF $A(a, bB)s$ reaction, used to transfer a cluster x off the particle a and populate excited states of nucleus F, later decaying to $b + B$.

Yet, astrophysical energies in the $A - x$ channel can be reached thanks to the energy paid to break the nucleus $a = x + s$. Such system, usually referred to as the Trojan Horse (TH) nucleus, has to show a large $x - s$ cluster structure. In a simplified picture, since the Coulomb and centrifugal barriers are overcome in the $A - a$ channel, the $A(x, b)B$ reaction takes place in the nuclear field so it is not hindered by Coulomb and centrifugal repulsion. In order to achieve a simple connection between the $A(a, bB)s$ cross section and the one of astrophysical interest ($A(x, b)B$), the quasi-free (QF) contribution to the $2 \to 3$ cross section has to be singled out, namely, a participant-spectator mechanism. This leads to the population of excited states of the intermediate nucleus $F = A + x$, later decaying into $b + B$, where x is coming from the virtual breakup of the TH nucleus a. An additional advantage is that the $x - s$ intercluster motion can be used to further reduce the $A - x$ relative energy (even reaching negative values by a careful choice of the kinematic variables[9]), as well as to span a broad $A - x$ relative energy interval even if a single beam energy is used, which is particularly useful in the case of radioactive ion beams [13, 14, 15] and neutrons [16].

In a graphical representation, the reaction mechanism sketched in Fig.1 has to be selected, namely, the participant (x)-spectator (s) process where the TH nucleus undergoes direct breakup and s is emitted without influencing the $A(x, b)B$ sub-reaction.

In the original theoretical formulations of the THM (see, for instance [17]), it was pointed out that for each contributing partial wave to the $A(x, b)B$ virtual reaction, the corresponding penetration factor $P_l(k_{xA}, R_{xA})$ (k_{xA} and R_{xA} being the $A - x$ channel wave number and radius, respectively) and normalization constant $N_{l,\,Ax}$ have to be multiplied to the $A(x, b)B$ cross section, obtained using the THM, to compare it with the directly measured one:

$$\frac{d\sigma}{d\Omega} = \sum_l N_{l,\,Ax} P_l(k_{xA}, R_{xA}) \frac{d\sigma^{HOES}_{Ax \to bB}}{d\Omega} \,. \tag{1}$$

In this formula, $d\sigma^{HOES}_{Ax \to bB}/d\Omega$ is the half-off-energy-shell (HOES) THM cross section, which is deduced from the THM data, since the participant x is a virtual particle that does not satisfy the mass-shell equation [22].

Of course, in the case of multi resonant reactions, such an approach may introduce large normalization errors, since it is necessary to determine a scaling constant for each contributing wave [17]. When many resonances are present, populated in a variety of orbital angular momenta, an extension of the THM to account for multiple wave contribution and normalization to the direct data is mandatory. After the seminal work [18], extensive theoretical [19, 20] and experimental [21, 22] studies were performed to develop and validate a novel approach to treat multi-resonance reactions, named modified R-matrix approach. This is a generalisation of the R-matrix [23] because we consider reactions with three particles in the exit channel, where the

IOP Conf. Series: Journal of Physics: Conf. Series **1308** (2019) 012012 doi:10.1088/1742-6596/1308/1/012012

TH-nucleus a in the initial state carries the transferred particle x, which is off-the-energy-shell.

In the plane wave approximation, the amplitude of the process $a + A \to b + B + s$ (Fig.1) is:

$$M^{PWA(prior)}(P, \mathbf{k}_{aA}) = \left\langle \chi_{sF}^{(0)} \Psi_{bB}^{(-)} | V_{xA} | \varphi_a \varphi_A \chi_{aA}^{(0)} \right\rangle , \tag{2}$$

where $P = (\mathbf{k}_{sF}, \mathbf{k}_{bB})$ is the six-dimensional momentum describing the three-body system s, b and B. $\chi_{aA}^{(0)} = \exp(i\mathbf{k}_{aA} \cdot \mathbf{r}_{aA})$, $\chi_{sF}^{(0)} = \exp(i\mathbf{k}_{sF} \cdot \mathbf{r}_{sF})$, $\mathbf{r}_{ij}$ and $\mathbf{k}_{ij}$ are the relative coordinate and relative momentum of i and j nuclei, $\Psi_{bB}^{(-)}$ is the wave function of the nuclei b and B in the exit channel, $F = b + B$, V_{xA} is the interaction potential of x and the target nucleus A, φ_a and φ_A are the a and A bound state wave functions, respectively. Assuming that the reaction yield is dominated by resonances, the plane wave amplitude from which the THM cross section is calculated is given by Refs. [24, 20]:

$$\begin{aligned}
M^{PWA(prior)}(P, \mathbf{k}_{aA}) &= (2\pi)^2 \sqrt{\frac{1}{\mu_{bB} k_{bB}}} \varphi_a(\mathbf{p}_{sx}) \\
\times \sum_{J_F M_F j' l l' m_{j'} m_l m_{l'} M_x} & i^{l+l'} \langle j m_j l m_l | J_F M_F \rangle \langle j' m_{j'} l' m_{l'} | J_F M_F \rangle \\
& \times \langle J_x M_x J_A M_A | j' m_{j'} \rangle \langle J_s M_s J_x M_x | J_a M_a \rangle \\
& \times \exp\left[-i\delta_{bB\,l}^{hs}\right] Y_{l m_l}(-\hat{\mathbf{k}}_{bB}) \\
& \times \sum_{\nu\tau=1}^{N} \left[\Gamma_{\nu bBjlJ_F}\right]^{1/2} \left[\mathbf{A}^{-1}\right]_{\nu\tau} Y_{l'm'}^{*}(\hat{\mathbf{p}}_{xA}) \\
& \times \sqrt{\frac{R_{xA}}{\mu_{xA}}} \left[\Gamma_{\nu xAl'j'J_F}(E_{xA})\right]^{1/2} P_{l'}^{-1/2}(k_{xA}, R_{xA}) \\
\times \Big(j_{l'}(p_{xA}R_{xA}) & \left[(B_{xA\,l'}(k_{xA}, R_{xA}) - 1) - D_{xA\,l'}(p_{xA}, R_{xA})\right] \\
+2Z_x Z_A e^2 \mu_{xA} & \int_{R_{xA}}^{\infty} \mathrm{d}r_{xA} \frac{O_{l'}(k_{xA}, r_{xA})}{O_{l'}(k_{xA}, R_{xA})} j_{l'}(p_{xA} r_{xA}) \Big) .
\end{aligned} \tag{3}$$

Here, $F = b + B$, μ_{ij} is the $i - j$ reduced mass, r_{ij} is the $i - j$ relative distance, $\mathbf{p}_{ij}$ is the $i - j$ relative momentum in the case of off-energy-shell particles, thus $E_{ij} \neq p_{ij}^2/2\mu_{ij}$ (while $\mathbf{k}_{ij}$ is calculated assuming the particles on-shell), $\delta_{bB\,l}^{hs}$ is the solid sphere scattering phase shift, R_{xA} the $x + A$ channel radius,

$$B_{xA\,l'}(k_{xA}, R_{xA}) = R_{xA} \frac{\frac{\partial O_{l'}(k_{xA}, R_{xA})}{\partial r_{xA}}|_{r_{xA}=R_{xA}}}{O_{l'}(k_{xA}, R_{xA})} \tag{4}$$

is the logarithmic derivative as in the R-matrix method,

$$O_{l'}(k_{xA}, R_{xA}) = \sqrt{\frac{k_{xA} R_{xA}}{P_{l'}(k_{xA}, R_{xA})}} \exp\left[-i\delta_{xA\,l'}^{hs}\right] \tag{5}$$

is the outgoing spherical wave,

$$D_{xA\,l'}(p_{xA}, R_{xA}) = R_{xA} \frac{\frac{\partial j_{l'}(p_{xA}, R_{xA})}{\partial r_{xA}}|_{r_{xA}=R_{xA}}}{j_{l'}(p_{xA}, R_{xA})} \tag{6}$$

is the logarithmic derivative and $j_{l'}(p_{xA}, R_{xA})$ the spherical Bessel function, N the number of the included levels.

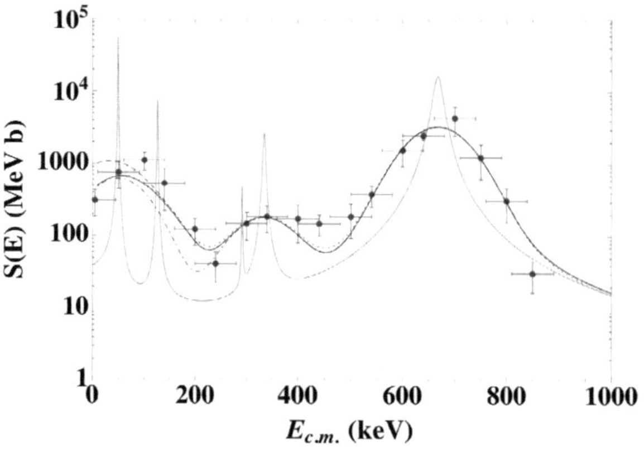

Figure 2. R-matrix analysis of the THM astrophysical factor (blu points), under the assumption of $J^\pi = 3/2^+$ for the 6460 keV ^{19}Ne state [14]. The solid black line is the smoothed R-matrix calculation, accounting for a 53 keV energy spread (standard deviation). The red line is the corresponding deconvoluted astrophysical factor. The dashed black line is the smoothed R-matrix calculation including the 6417 keV level, while the dot-dashed line is the smoothed R-matrix calculation where the 6537 keV is excluded. Finally, the dotted line marks the smoothed R-matrix calculation where the interference signs were changed to (++)(-+). The figure is taken from [15].

Eq.3 contains the same level matrix $\mathbf{A}_{\nu\tau}$ as in the standard R-matrix theory [23], thus the same energies and reduced width amplitudes. Therefore, from the HOES cross section it is possible to deduce the on-energy-shell (OES) one with no need of extrapolation, accounting for the virtual nature of the transferred particle and energy resolution effects. Also, in the plane wave approximation the OES cross section is obtained in arbitrary units; however, normalization to direct data is gained by scaling the reduced γ-widths of indirectly measured resonances to the ones of states lying at high energy and accurately known from the literature, independently from the corresponding orbital angular momenta (see [25]). Another important feature is the possibility to measure down to zero energy, even in the case of charged-particle induced reactions, which is accomplished thanks to the appearance of the $P_{l'}^{-1/2}(k_{xA}, R_{xA})$ factor in Eq.3, namely the inverse penetration factor.

3. The ^{18}F$(p,\alpha)^{15}$O reaction studied with the THM

Crucial information on nova nucleosynthesis can be potentially inferred from γ-ray signals powered by ^{18}F decay [26]. The ^{18}F$(p,\alpha)^{15}$O cross section has been measured by means of several dedicated experiments, being a major ^{18}F destruction channel. However, the present uncertainty on the cross section at astrophysical energies, below about $\sim$ 300 keV is far from satisfactory, owing to the need of radioactive ^{18}F beam. Thanks to the suppression of the Coulomb barrier, the THM proved very well suited to span the whole energy interval in a single measurement [13, 14]. Then, following [27], we performed a R-matrix analysis of the THM data to check the compatibility of the THM S-factor with this recent R-matrix calculation, accounting for most of recent direct and indirect results and spectroscopic information.

The THM S(E) factor is shown as blue circles in Fig.2. The best fit of the THM data [13, 14] was achieved by assuming the (++)(++) interference pattern (using the same notation as in [27]), by excluding the 7 keV resonance, due to the $3/2^-$ state of ^{19}Ne at 6417 keV energy, and by introducing the $7/2^+$ state of ^{19}Ne at 6537 keV, whose occurrence was pointed out in [13, 14]. In detail, the (++)(++) notation is used when the relative interference signs between the -124 keV

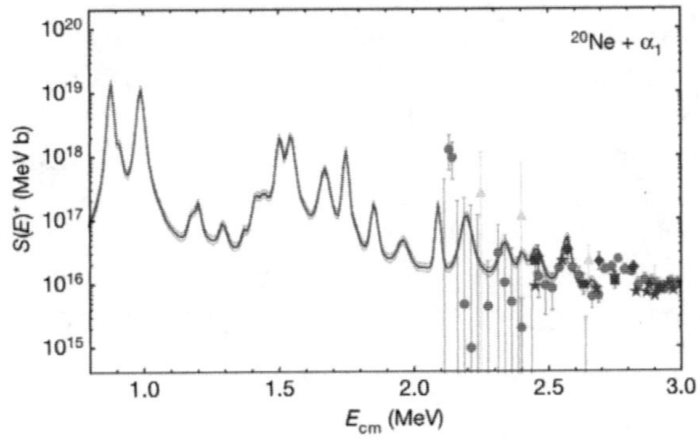

Figure 3. THM $^{12}C(^{12}C, \alpha_1)^{20}Ne$ modified astrophysical factor S^* (gray band), compared with the available direct data (solid symbols). See Ref.[33] for a detailed discussion.

and 1461 keV resonances in the $^{18}F(p, \alpha)^{15}O$ cross section is (++) (first pair) and the ones of the 47 and 665 keV resonances is (++) as well (second pair). The list of levels used in the calculations is given in Tab.1 of [15], while the resulting best fit astrophysical factor is marked by a red line in Fig.2 in the case of infinite energy resolution, and by a black solid line including the effect of experimental energy resolution (53 keV standard deviation, [13]). For 18 degrees of freedom, a reduced $\chi^2 = 1.5$ is obtained. The present result is not much sensitive on the change of the interference pattern from the (++)(++) relative signs to the (++)(-+) combination, that is, assuming constructive interference between the $3/2^+$ states (dotted line in Fig.2), thanks to the occurrence of the 126 keV peak. Instead, the inclusion of the 6417 keV state (dashed black line) or the removal of the 6537 keV one (dot-dashed black line) cause a significant increase of the reduced χ^2, suggesting that such alternative solutions should be discarded.

4. The $^{12}C + ^{12}C$ reactions

$^{12}C + ^{12}C$ fusion is a key process in nuclear physics and astrophysics. Indeed, it has been extensively investigated (see Ref. [28] and references therein) to understand fusion reaction mechanism, especially to pinpoint the occurrence and origin of the fusion hindrance (see, for instance, Ref.[29]). In astrophysics, the $^{12}C + ^{12}C$ fusion rate bears a great importance in the evolution of massive stars ($M_\odot \geq 8$), especially in the case of objects in late evolutionary stages. At present, direct data are present down to about 2.5 MeV center-of-mass energies (at lower energies uncertainties are so large that they are not statistically significant [28]), so extrapolation is necessary to supply the astrophysical factor at the energies of interest, especially for astrophysics, comprised between about 1–2 MeV. Extrapolations are usually performed using the modified astrophysical S-factor [30]:

$$S^* = E\,\sigma(E)\exp(2\pi\eta + g\,E)\,, \tag{7}$$

where $\eta = 13.88\,E^{-1/2}$ and $g = 0.46$ MeV^{-1} (the center-of-mass energy E being expressed in MeV).

However, the large uncertainties affecting direct data close to astrophysical energies and the presence of oscillations in the modified astrophysical factor make extrapolation quite unreliable, especially because the reaction mechanism (for instance, the possibility that fusion hindrance suppresses the cross section to values lower than predicted by coupled-channel calculations [29]). In the case of $^{12}C+^{12}C$, measurements below 2 MeV are hampered because the Coulomb barrier reduces the cross section to values as low as few nanobarns already at $E = 2.5$ MeV, because of the strong background especially due to the deuterium content in the targets [28] and because of the effect of energy loss in the targets themselves, making it difficult to fix the correct interaction energy [31]. For all these reasons, we decided to carry out a THM measurement, which can bypass all the mentioned problems thanks to its indirect approach. Indeed, while the Coulomb suppression is overcome as discussed in Sect.2, the use of reactions with three particles in the exit channel makes it possible to reconstruct the reaction Q-value and select only the events from the reaction of interest, essentially background free. Also, the use of comparatively higher energies than the astrophysical ones make the ambiguity in interaction energy determination negligible.

Since the possibility of using ^{14}N to transfer a ^{12}C nucleus is well known (see, for instance, [32]), we investigated ^{12}C + ^{12}C fusion by applying the THM to the ^{14}N + ^{12}C reactions [33]. The experiment was carried out at the *Laboratori Nazionali del Sud, Istituto Nazionale di Fisica Nucleare* (Catania, Italy) using a 30 MeV ^{14}N beam delivered by the Van de Graff Tandem accelerator. We were able to measure the cross sections for the ^{12}C(^{12}C, α_0)^{20}Ne, ^{12}C(^{12}C, α_1)^{20}Ne, ^{12}C(^{12}C, p_0)^{23}Na and ^{12}C(^{12}C, p_1)^{23}Na channels in arbitrary units, covering the energy range of astrophysical interest and a region above about 2.5 MeV. This is necessary to normalize the THM data to the direct ones, so an energy interval where they are less affected by statistical and systematic errors is chosen, to obtain the THM S^* factors and minimize systematics errors. The THM measurement made it possible to perform a validity test of the method, thanks to the precise matching of the trend of the experimental direct and indirect data in the overlapping energy region, and to shed light on the trend of S^* below about 2 MeV, where a definite resonant pattern was observed, whose interesting physical consequences are being addressed by several authors [34, 35]. In particular, fig.3 from [33] shows the ^{12}C(^{12}C, α_1)^{20}Ne S^*-factor obtained by applying Eq.3, superimposed to the available direct data (filled circles[36], filled squares[37], empty diamonds[38], filled stars[39] and filled triangles[40]). Other approaches, such as those sketched in Refs. [41, 42], presently fail to satisfy such necessary condition, as shown in Ref. [43], demonstrating once again the robustness of THM.

References

[1] Broggini C et al. 2018 *Progress in Particle and Nuclear Physics* **98** 55
[2] Iliadis C 2007 *Nuclear Physics of Stars* (New York: Wiley)
[3] Rolfs C E and Rodney W S 1998 *Cauldrons in the cosmos: Nuclear astrophysics* (Chicago: University of Chicago Press)
[4] Bracci L et al. 1990 *Nucl. Phys. A* **513** 316
[5] La Cognata M et al. 2005 *Phys. Rev. C* **72** 065802
[6] Rinollo A et al. 2005 *Nucl. Phys. A* **758** 146C
[7] Tumino A et al. 2011 *Phys. Lett. B* **705** 546
[8] La Cognata M et al. 2012 *Phys. Rev. Lett.* **109** 232701
[9] La Cognata M et al. 2013 *The Astrophysical Journal* **777** 143
[10] Trippella O and La Cognata M 2017 *The Astrophysical Journal* **837** 41
[11] Spitaleri C 1991 *Proceedings of the 5Th Winter School on Hadronic Physics: Problems of Fundamental Modern Physics II* (Singapore: World Scientific) 21
[12] Spitaleri C et al. 2016 *Eur. Phys. J. A* **52** 77
[13] Cherubini S et al. 2015 *Phys. Rev. C* **92** 015805
[14] Pizzone R G et al. 2016 *Eur. Phys. J. A* **52** 24
[15] La Cognata M 2017 *The Astrophysical Journal* **846** 65
[16] Gulino M et al. 2013 *Phys. Rev. C* **87** 012801
[17] Typel S et al. 2003 *Annals of Physics* **305** 228
[18] La Cognata M 2008 *Phys. Rev. Lett.* **101** 152501
[19] Mukhamedzhanov A M et al. 2008 *Journal of Physics G Nuclear Physics* **35** 014016
[20] Mukhamedzhanov A M 2011 *Phys. Rev. C* **84** 044616
[21] La Cognata M et al. 2010 *Astrophys. J.* **723** 1512
[22] La Cognata M et al. 2009 *Phys. Rev. C* **80** 012801

[23] Lane A M and Thomas R G 1958 *Rev. Mod. Phys.* **30** 257
[24] Tribble R E et al. 2014 *Reports on Progress in Physics* **77** 106901
[25] La Cognata M et al. 2015 *Astrophys. J.* **805** 128
[26] José, J 2016 *Stellar Explosions: Hydrodynamics and Nucleosynthesis* (Boca Raton, FL, London: CRC/Taylor and Francis)
[27] Bardayan D W et al. 2015*Phys. Lett. B* **751** 311
[28] Zickefoose J et al. 2018 *Phys. Rev.* C **97** 065806
[29] Jiang C L et al. 2002 *Phys. Rev. Lett.* **89** 052701
[30] Patterson J R et al. 1969 *The Astrophysical Journal* **157** 367
[31] Fisichella M et al. 2015 *Phys. Rev.* C **92** 064611
[32] Artemov K P et al. 1984 *Phys. Lett. B* **149** 325
[33] Tumino A et al. 2018 *Nature* **557** 687
[34] Mori K et al. 2019 *Monthly Notices of the Royal Astronomical Society: Letters* **482** L70
[35] Straniero O et al. 2019 arXiv:1901.00173
[36] Spillane T et al. 2007 *Phys.Rev. Lett.* **98** 122501
[37] Mazarakis M G et al. 1973 *Phys. Rev.* C **7** 1280
[38] High M D et al. 1977 *Nucl. Phys.* A **282** 181
[39] Kettner K U et al. 1980 *Z. Phys.* A **298** 65
[40] Barron-Palos L et al. 2006 *Nucl. Phys.* A **779** 318
[41] Mukhamedzhanov A M and Pang D Y 2018 arXiv:1806.08828v4
[42] Mukhamedzhanov A M et al. 2018 arXiv:1806.05921v1
[43] Tumino A et al. 2018 arXiv:1807.06148v1

IOP Conf. Series: Journal of Physics: Conf. Series **1308** (2019) 012013 doi:10.1088/1742-6596/1308/1/012013

Analysis of states in ^{13}C and their cluster structure

I. Lombardo

Istituto Nazionale di Fisica Nucleare - Sezione di Catania, Via S. Sofia, I-95125 Catania, Italy

E-mail: ivano.lombardo@ct.infn.it

Abstract. Accurate studies on ^{13}C spectroscopy have great impact in the present understanding of the role played by extra-neutrons in stabilizing α-cluster structures formed in light nuclei. Carbon-13 is in fact the simplest systems that can be formed by adding a neutron to make a triple-α molecular-like structure. The accurate spectroscopy of excited states at energies above the α emission threshold is fundamental to benchmark the calculations of theoretical models aiming to describe clustering in light nuclei.

To improve our knowledge of ^{13}C structure, we performed a comprehensive R-matrix fit of α+^{9}Be elastic and inelastic scattering data in the energy region $E_\alpha \approx$ 3.510 MeV at several backward polar angles. To carefully determine the partial decay widths of states above the α-decay threshold we included in the fit also ^{9}Be$(\alpha, n)^{12}$C cross section data taken from the literature. This analysis allows to improve the (poorly known) spectroscopy of excited states in ^{13}C in the $E_x \approx$ 12-17 MeV region, and tentatively suggests the presence of a large-deformation negative-parity molecular band.

1. Introduction

The structure of light nuclear systems is very special and unusual. The dominance of surface effects can lead to a rearrangement between neutrons and protons to form α-like clusters. They continuously (and dynamically) dissolve and clot in the nuclear medium, leading to important deviation from the spherical symmetry and introducing new types of symmetries in the nuclear domain [1, 2]. This behaviour is typical of the so called self-conjugate nuclei, that are even-even nuclei with an equal number of neutrons and protons. Beryllium-8, ^{12}C, ^{16}O, ^{20}Ne are the most representative cases; their α-cluster nature emerges both in the structure of ground state (e.g. in the case of ^{8}Be [3]) or of excited states near and above the α-decay threshold. An example of the latter case is the famous Hoyle state of ^{12}C: from the '50 to recent times, it was the subject of several theoretical and experimental studies [4, 5].

The cluster structure of nuclei plays a fundamental role also in collateral domains of nuclear physics. For example, in nuclear astrophysics, the cluster structure of self-conjugate nuclei such as ^{12}C and ^{20}Ne, or of non-self-conjugate systems such as ^{11}C, above particle-emission threshold, has an influence on reaction rates of stellar interest (see, for example, Refs. [6–10]). Furthermore, in collisions involving heavy ions at intermediate energies (e.g. Refs. [11–13]), the nuclear system can reach a dilute density condition in which alpha clusters can be copiously produced, and this process can have an influence on the dynamical evolution and the disassembly of the transient system consequently formed [14–16].

In the presence of extra-neutrons, new manifestations of α-clustering can be observed [17]. In analogy with physical chemistry, extra-neutrons can give rise to a *covalent* bonding between alpha-centres. This behaviour has been experimentally ascertained for example, in ^{9}Be, where

IOP Conf. Series: Journal of Physics: Conf. Series **1308** (2019) 012013 doi:10.1088/1742-6596/1308/1/012013

the two α-centres are bound together by the valence neutron, leading to a super-deformation even in the ground state; on top of this, rotational states have been discovered [17]. Similar discussions apply to other neutron-rich beryllium isotopes (e.g. ^{10,11,12}Be) at high excitation energies [18].

For nuclear states characterized by three centres, the phenomenology of clustering is still unclear. Several microscopic models pointed out the existence of a molecular-like structures of neutron rich carbon isotopes ^{13,14,16}C at high excitation energies [19, 20], but, unfortunately, experimental data on such aspects are incomplete or even missing [21–25]. Within this framework, the analysis of states of the simplest triple-α neutron-rich system, ^{13}C, is a key ingredient. For such a nucleus, theoretical calculations suggest the appearance of two molecular bands with opposite parities, $K^{\pi} = 3/2^{\pm}$, in an excitation energy region close to the $3\alpha + n$ disintegration energy at 12.22 MeV [19, 26]. It is very interesting to underline that, in this conference, new, very interesting theoretical predictions on the structure of ^{13}C exploiting the algebraic symmetries of the Cluster Shell Model (CSM) were discussed.

Despite a large theoretical effort, the literature still shows too many ambiguities concerning the experimental J^{π} values of states in ^{13}C in the excitation energy range relevant for clustering studies (i.e., $E_x \approx 13-16$MeV) [27]. Because of such ambiguities, a clear picture of the structure of ^{13}C is still missing and no well grounded conclusions on the existence of molecular bands in ^{13}C can be drawn. To clarify these aspects, we performed a new experiment on α+^{9}Be resonant elastic scattering data with the aim of improving the spectroscopy of high energy states in ^{13}C. According to the literature, this reaction channel seems to be highly selective for cluster states in ^{13}C because of the pronounced cluster structure of the ^{9}Be target nucleus.

The resonance parameters (partial widths, J^{π}, interference signs, resonance energies, etc) of excited states in the domain $E_x \approx 11.8 - 18$ MeV were estimated by a comprehensive R-matrix fit of cross section data by using the AZURE2 code [28, 29] of various reaction channels: the resonant scattering and the ^{9}Be$(\alpha, n)^{12}$C reactions. The detailed discussion on such results is given in the more extended Ref. [30]; in this proceedings, we will discuss some relevant results obtained from the data analysis.

2. R-matrix analysis on experimental data

The database used for the present analysis involves excitation functions of the α+^{9}Be elastic scattering derived from an experiment performed at the TTT3 tandem accelerator in Naples, Italy [31–33]. To obtain the excitation functions, the beam energy was finely changed from 3.3 MeV up to 10 MeV, in steps of 60 keV. The target was a thin self-supporting ^{9}Be foil, having a thickness of 122 $\mu g/cm^2$ and some traces of C and O contaminants. The detection device constituted a system of silicon detectors placed at several angles in the backward hemisphere. More experimental details on such apparatus are discussed in Refs. [30, 34, 35]. From the available data, it was possible to measure excitation functions for both elastic and inelastic (to the first excited state in ^{9}Be) scattering channels. The absolute cross section values derived with the thin target experiment was checked by means of a thick target experiment [35], giving good consistency to the data. Values of the absolute cross sections for the ^{9}Be$(\alpha, n)^{12}$C reactions were taken from the literature ([36, 37]); the need for a normalization factor between the data sets reported in the literature is detailed in Ref. [30].

The initial parameters of excited states in ^{13}C used in terms of the R-matrix to analyse the experimental data were derived from the table of states of Ref. [27]; we took into account also recent findings reported in Refs. [21, 38]. When contrasting assignments were reported in the literature, the J^{π} values were chosen in order to reproduce, in the best possible way, the trend of excitations functions data. Finally, we had success in reproducing quite well all the experimental data with the R-matrix fit, as we show in the Figures of Ref. [30]. As an example, in Figure 1 of this proceedings we show a part of the differential cross section for the elastic scattering

IOP Conf. Series: Journal of Physics: Conf. Series **1308** (2019) 012013 doi:10.1088/1742-6596/1308/1/012013

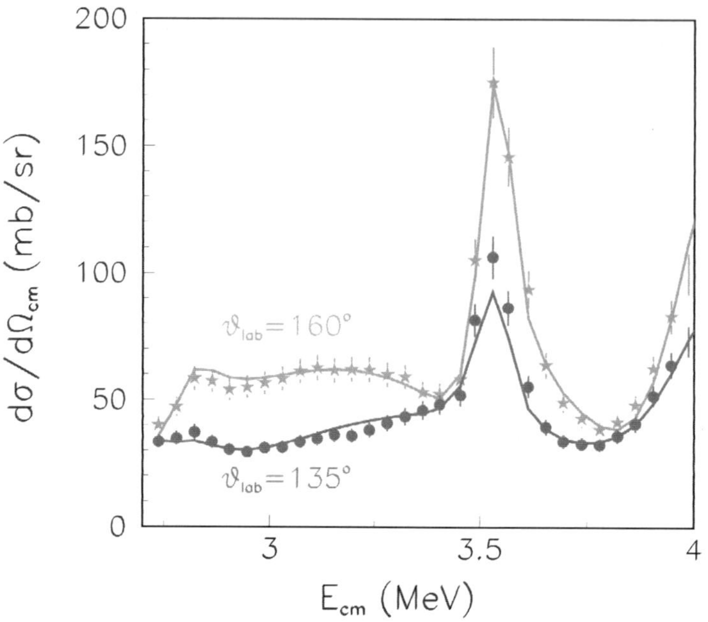

Figure 1. Excitation functions measured at $\theta_{lab} = 160°, 135°$ (green solid stars and brown circles, respectively) for the α+^{9}Be elastic scattering in the E_{cm} =2.7-4 MeV range. The solid green and brown lines represent the R-matrix fit of experimental data.

channel at $\theta_{lab} = 160°, 135°$, together with the R-matrix fit (solid lines); they nicely agree.

We can derive some interesting considerations from the R-matrix analysis performed. For example, the small structure seen at $E_{cm} \approx 2.8$ MeV has been associated with a high-spin ($J^\pi = 9/2^-$) state in ^{13}C; this state has a sizeable reduced partial width if compared with the Wigner limit for the α+^{9}Be channel [30]. As we will discuss later, this state is a candidate to be part to the $K^\pi = 3/2^-$ negative parity molecular band in ^{13}C.

At higher energies, the pronounced maximum seen at $E_{cm} \approx 3.5$ MeV has been associated in the past with a resonance in the compound nucleus with contrasting J^π assignments; furthermore, it was included (as $J^\pi = 9/2^-$) in the systematics of molecular cluster states reported in Ref. [19]. The presence of a benchmarked data set of elastic scattering excitation functions, and the availability of data at several angles allow us to firmly assign the $J^\pi = 5/2^-$ value to such resonance, due to the 14.13 MeV excited state. The alternative $J^\pi = 9/2^-$ assignment suggested in Ref. [19] does not reproduce the experimental data, both in term of the shape and the intensity of the peak seen in this energy region. As shown in Figure 1, a good reproduction of elastic scattering data at two different angles is obtained when we assume a $5/2^-$ assignment for the 14.13 MeV excited state ($E_{cm} \approx 3.5$ MeV peak), a $9/2^-$ assignment for the 13.41 MeV excited state ($E_{cm} \approx 2.8$ MeV peak), and the presence of two broad states (with $7/2^-$ and $5/2^+$ assignments) at 13.49 and 13.63 MeV; the last two states explain the trend of elastic scattering data in the $E_{cm} \approx 3 - 3.5$ MeV region.

In Ref. [30] we report the full discussion for all the states included in the level scheme of ^{13}C above the alpha emission threshold, and Table 1 lists the resonance parameters obtained from the present analysis. In particular, the coupling of a sub-alpha-threshold state at 9.897 MeV ($J^\pi = 3/2^-$), a state at 10.818 MeV ($J^\pi = 5/2^-$) very close to the threshold (at energies lower than the ones explored in the present analysis) and the two states at 12.45 MeV ($J^\pi = 5/2^-$)

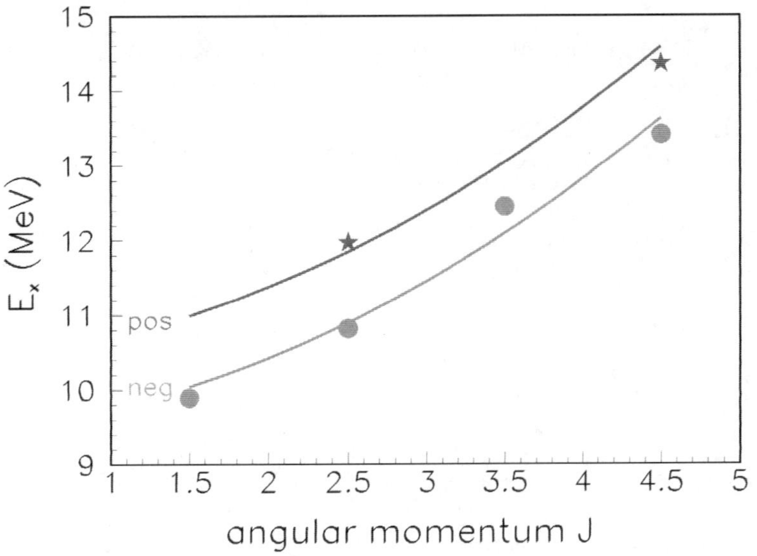

Figure 2. The $K^\pi = 3/2^\pm$ rotational bands suggested from the spectroscopy of excited states reported in the literature, complemented with the data obtained from the present R-matrix analysis. The solid lines represent simple calculations of molecular rotational bands for a structure with $\frac{\hbar^2}{2\Im} \approx 0.17$ MeV moment of inertia parameter.

and 13.41 MeV ($J^\pi = 9/2^-$) seems to suggest the presence of a rotation band with negative parity, as visible in Figure 2 (solid circles). A simple calculation assuming a moment of inertia parameter of $\frac{\hbar^2}{2\Im} \approx 0.17$ MeV describes the experimental data well; the value of the $\frac{\hbar^2}{2\Im}$ parameter is quite similar to the one determined theoretically from microscopic models in which the carbon has a very pronounced molecular nature [39], and stimulates further analysis on this aspect. Two positive parity states at 11.97 ($J^\pi = 5/2^+$) and 14.36 MeV ($J^\pi = 9/2^+$) seem to belong to a positive parity band with features similar to the negative parity one, but the situation in this case is less clear and calls for new experiments.

3. Conclusions

In this proceeding, we discuss some findings obtained by investigating ^{13}C spectroscopy in the excitation energy domain going from 11.8 up to about 18 MeV. This excitation energy region is very interesting because the existence of two opposite parity molecular bands with cluster nature has been suggested by various theoretical calculations. To improve our knowledge on this topic, we performed a comprehensive R-matrix analysis of various reaction data sets populating ^{13}C excited states in the compound nucleus scenario. As a result, it was possible to improve several J^π assignments for some excited states for which tentative values were reported in the literature. The possible existence of a negative parity molecular band with a moment of inertia compatible with a molecular-like structure in ^{13}C has also been suggested from the analysis of excited states. As a perspective, we believe that in the future, the availability of new radioactive ion beam facilities able to identify the beam particles one by one [40], and the use of the high-granularity and/or large solid angles arrays for light particles [41–43] and/or gamma and neutron identification [44–46], would be very useful to unveil the cluster structure of neutron-rich carbon isotopes.

References

[1] Marin-Lambarri D J, Bijker R, Freer M, Gai M, Kokalova T, Parker D J and Wheldon C 2014 *Phys. Rev. Lett.* **113** 012502
[2] Bijker R and Iachello F 2000 *Phys. Rev.* **C61** 067305
[3] Wiringa R B *et al.* 2000 *Phys. Rev. C* **62** 014001
[4] Dell'Aquila D *et al.* 2017 *Phys. Rev. Lett.* **119** 132501
[5] Smith R *et al.* 2017 *Phys. Rev. Lett.* **119**(13) 132502
[6] Buchmann L R and Barnes C A 2006 *Nucl. Phys. A* **777** 254–290
[7] Cognata M L *et al.* 2015 *Astrophys. J.* **805** 128
[8] Lombardo I *et al.* 2015 *Phys. Lett. B* **748** 178
[9] He J J *et al.* 2018 *Chin. Phys. C* **42** 15001
[10] Spitaleri C *et al.* 2017 *Phys. Rev.* **C95** 035801
[11] Russotto P *et al.* 2015 *Phys. Rev. C* **91** 014610
[12] Filippo E D *et al.* 2009 *Acta Phys. Pol. B* **40** 1199
[13] Lombardo I *et al.* 2010 *Nucl. Phys. A* **834** 458–460c
[14] Schmidt K *et al.* 2017 *Phys. Rev. C* **95** 054618 (*Preprint* 1612.00267)
[15] Borderie B *et al.* 2016 *Phys. Lett. B* **755** 475–480
[16] Marini P *et al.* 2016 *Physics Letters B* **756** 194–199
[17] von Oertzen W, Freer M and Kanada-En'yo Y 2006 *Phys. Rep.* **432** 43–113
[18] Suzuki D *et al.* 2013 *Phys. Rev. C* **87** 054301
[19] Milin M and von Oertzen W 2002 *Eur. Phys. J. A* **14** 295–307
[20] Baba T, Chiba Y and Kimura M 2014 *Phys. Rev.* **C 90** 064319
[21] Wheldon C *et al.* 2012 *Phys. Rev. C* **86** 044328
[22] Freer M *et al.* 2014 *Phys. Rev. C* **90** 054324
[23] Dell'Aquila D *et al.* 2016 *Phys. Rev. C* **93** 024611
[24] Dell'Aquila D 2016 *Il Nuov. Cim. C* **39** 272
[25] Dell'Aquila D *et al.* 2016 *EPJ Web of Conf.* **117** 06011
[26] Furutachi N and Kimura M 2011 *Phys. Rev. C* **83** 021303(R)
[27] Ajzenberg-Selove F 1991 *Nucl. Phys. A* **1** 523
[28] Azuma R E *et al.* 2010 *Phys. Rev.* **C81** 045805
[29] Wiescher M *et al.* 2017 *Phys. Rev. C* **95** 044617
[30] Lombardo I, Dell'Aquila D, Spadaccini G, Verde G and Vigilante M 2018 *Phys. Rev. C* **97**(3) 034320 URL https://link.aps.org/doi/10.1103/PhysRevC.97.034320
[31] Lombardo I *et al.* 2013 *J. Phys. G* **40** 125102
[32] Lombardo I *et al.* 2014 *Bulletin of the Russian Academy of Sciences: Physics* **78** 1093
[33] Lombardo I *et al.* 2016 *J. Phys. G.* **43** 45109
[34] Lombardo I *et al.* 2014 *J. Phys. Conf. Ser.* **569** 012068
[35] Lombardo I *et al.* 2013 *Nucl. Instrum. Meth. Phys. Res.* **B 302** 19–23
[36] Geiger K W *et al.* 1975 *Nuclear Instruments and Methods* **131** 315–321
[37] Wrean P R, Brune C R and Kavanagh R W 1994 *Phys. Rev. C* **49** 1205–1213
[38] Freer M *et al.* 2011 *Phys. Rev.* **C 84** 034317
[39] Furutachi N and Kimura M 2011 *Phys. Rev.* **C83** 021303
[40] Lombardo I *et al.* 2011 *Nucl. Phys. B Proc. Suppl.* **215** 272–274
[41] Verde G *et al.* 2013 *J. Phys. Conf. Ser.* **420** 0112158
[42] Pagano E V *et al.* 2016 *EPJ Web of Conf.* **117** 10008
[43] Dell'Aquila D *et al.* 2018 *Nucl. Instrum. Meth. Phys. Res. A* **877** 227
[44] Cardella G *et al.* 2015 *Nucl. Instr. and Meth. A* **799** 64
[45] Modamio V *et al.* 2015 *Nucl. Instr. Meth. Phys. Res. A* **775** 71
[46] Martorana N S *et al.* 2018 *Physics Letters B* **782** 112–116

LLNL-PROC-769943

Microscopically based energy density functionals for nuclei using the density matrix expansion

R. Navarro Pérez[1,2], **N. Schunck**[2]

[1] Department of Physics, San Diego State University, San Diego, California 92182, USA
[2] Nuclear and Chemical Sciences Division, Lawrence Livermore National Laboratory, Livermore, CA 94551, USA

E-mail: rnavarroperez@sdsu.edu

Abstract. While ab initio many-body techniques have been able to successfully describe the properties of light and intermediate mass nuclei based on chiral effective field theory interactions, neutron-rich nuclei still remain out of reach for these methods. Conversely, energy density functional approaches can be used to calculate properties of heavy nuclei but rely mostly on phenomenological interactions. A usable form of the nuclear energy density functional that is rooted in the modern theory of nuclear forces was presented recently. The first component of this new set of functionals corresponds to the direct part (Hartree term) of the expectation value of local chiral potentials on a Slater determinant. The exchange term, which is a functional of the non-local density, is transformed into a local functional by applying the density matrix expansion. In order to reduce the computational cost due to the direct implementation of non-separable, local interactions in the Hartree term, we use an approximation to represent the regularized Yukawa functions in terms of a sum of (separable) Gaussian functions. These proceedings analyze the accuracy of such an approximation in terms of the number of Gaussian functions and look for an optimal value that gives an acceptable level of accuracy while maintaining the computational memory requirements in a many-body calculation as low as possible.

1. Introduction

One of the overarching goals of nuclear physics is to find a theory capable of describing nuclear structure and reactions in terms of the underlying interactions between quarks and gluons. The description of the interaction between nucleons in Chiral Effective Field Theory (χ-EFT) provides the desired microscopic underpinning and also exhibits systematic, order-by-order, improvement [1–4]. Ab initio methods employ these realistic interactions and calculate properties of nuclear systems by directly considering the interaction between each pair of nucleons. This approach has been successful in describing light and medium nuclei (up to $Z = 50$) [5]. The same systematic, order-by-order, improvement from the interaction can be observed in the description of nuclear properties calculated with ab initio methods [6]. This allows keeping track of the predictive power of these type of theoretical calculations when looking at properties that have not been measured yet. However heavy systems, particularly neutron-rich systems, remain out of reach for ab initio methods. These neutron-rich systems are of great interest in view of upcoming measurements in flagship facilities like the Facility for Rare Isotope Beams (FRIB) and for the description of astrophysical phenomena like the r-process [7, 8].

In contrast to ab initio methods, Energy Density Functional (EDF) methods, also known as self-consistent mean-field approaches, substitute the interaction due to all pairs of nucleons by an "average" of the interaction with all the other nucleons [9, 10]. This substitution significantly reduces the computational cost of many-body nuclear calculations and allows estimating properties of heavy systems such as the neutron-rich nuclei mentioned earlier. However, the form of the interaction used in EDF approaches is usually constructed *ad hoc* in order to simplify calculations, and parameters are adjusted to reproduce experimental nuclear properties such as masses, charge radii or spin-orbit splittings. This leads to effective interactions of rather phenomenological nature whose predictive power, when extrapolated to properties beyond the ones used to determine its parameters, can be hard to maintain or even estimate.

Following earlier studies [11], we have recently proposed a new method to determine an EDF from microscopic nuclear potentials derived from χ-EFT through a tool known as the Density Matrix Expansion (DME). In practice, the new EDF looks like a phenomenological Skyrme-like contact interaction to which is added a more microscopic component which takes the form of density-dependent couplings resulting from applying the DME to the non-local one-body density [12]. This approach encodes the long-range physics contained in the χ-EFT potential. The phenomenological Skyrme-like part accounts for the short-range physics and its parameters need to be readjusted in order to reproduce experimental nuclear structure data.

The computational implementation of these semi-phenomenological interactions involves using certain numerical approximations (beyond the DME itself). In particular, the contribution of the long-range part of the chiral potential to the nuclear binding energy can be efficiently computed by approximating the Yukawa potentials with a sum of Gaussian functions. This approximation allows to recast the χ-EFT-derived potential in a form analogous to the purely phenomenological Gogny interaction, taking advantage of fact that this type of interaction is already implemented in several many-body Density Functional Theory (DFT) codes. Of course the implementation of any approximation requires to ensure a desired level of accuracy in order to avoid the introduction and propagation of numerical errors in all kinds of calculations. These proceedings analyze the convergence of the sum of Gaussian functions approximation to the central terms in the chiral potential at next to leading order (NLO) and next to next to leading order (N2LO). Section 2 gives a brief overview of the implementation of semi-phenomenological EDF introduced in [12]. Following that, section 3 presents an analysis of the accuracy of the Gaussian expansion depending on the number of terms included. Conclusions and outlook are discussed in section 4.

2. Semi-phenomenological EDF

Our many-body theoretical framework is the Hartree-Fock-Bogoliubov (HFB) formalism. In this framework, the many-body wave function is taken as a quasiparticle vacuum [9]. The total energy is expressed as a functional of the generalized density, or equivalently, as a functional of the one-body density matrix ρ and the pairing tensor κ. The full EDF can often be broken down into a sum of several terms, one that depends exclusively on ρ (the particle-hole channel) and another one that also involves κ and is only active when pairing correlations are present (the particle-particle channel).

The EDF in the particle-hole channel is often constructed by representing the many-body wave function by a Slater determinant and using the Wick theorem to express the energy due to many-body interactions as a functional of the density of particles in the system. In the simplest case of two-body interactions, this leads to writing

$$E^{\rm NN}[\rho] = \frac{1}{2}\sum_{ij}\langle ij \left| V^{\rm NN}(1 - P_\sigma P_\tau P_r)\right| kl\rangle \rho_{ki}\rho_{lj}, \tag{1}$$

with ρ_{ij} the matrix elements of the one-body density matrix on an arbitrary basis of the single-particle Hilbert space, P_σ, P_τ and P_r the usual spin-, isospin- and position-exchange operators and V^{NN} the two-body nuclear potential. The anti-symmetrization operator $(1 - P_\sigma P_\tau P_r)$ results in direct and exchange contributions, also known as the Hartree and Fock energies. After transforming the spatial coordinates to relative ($\boldsymbol{r}$) and center-of-mass ($\boldsymbol{R}$) coordinates, the Hartree and Fock energies become

$$E^{\mathrm{NN}}_{\mathrm{Hartree}} = \frac{1}{2}\mathrm{Tr}_1\mathrm{Tr}_2 \int \mathrm{dR} \int \mathrm{dr} \langle \mathrm{V}^{\mathrm{NN}}(\mathrm{r}) \rangle \rho_1 \left(\mathrm{R} + \frac{\mathrm{r}}{2}\right) \rho_2 \left(\mathrm{R} - \frac{\mathrm{r}}{2}\right) \tag{2}$$

$$E^{\mathrm{NN}}_{\mathrm{Fock}} = -\frac{1}{2}\mathrm{Tr}_1\mathrm{Tr}_2 \int \mathrm{dR} \int \mathrm{dr} \langle \mathrm{V}^{\mathrm{NN}}(\mathrm{r}) \rangle \rho_1 \left(\mathrm{R} - \frac{\mathrm{r}}{2}, \mathrm{R} + \frac{\mathrm{r}}{2}\right) \rho_2 \left(\mathrm{R} + \frac{\mathrm{r}}{2}, \mathrm{R} - \frac{\mathrm{r}}{2}\right) \mathrm{P}^{\sigma\tau}_{12}, \tag{3}$$

where the potential bracket includes spin and isospin states, the traces refer to summations over those quantum numbers and the local density matrix is $\rho(\boldsymbol{x}) = \rho(\boldsymbol{x}, \boldsymbol{x})$; for details on how to incorporate three-body forces into this formalism, see refs. [12, 13]. In a purely phenomenological approach the spatial form of the two-body interaction V^{NN} is partly chosen to simplify the calculation of these energy terms. Popular choices include delta functions for Skyrme functionals [14, 15] and Gaussian functions for the Gogny interaction [16]. Note that by representing the many-body wave function with a Slater determinant, the EDF approach assumes independent particles. Many-body correlations are recovered by a combination of two mechanisms: the spontaneous breaking of symmetries of the nuclear potential and the calibration of the parameters of the energy-density dependent functional to experimental data [9].

In order to anchor the EDF approach to a more microscopic theory and a systematic, order-by-order, description of nuclear forces, the new microscopic functionals presented in [12] start from the local chiral potentials up to N2LO with and without Δ excitations as outlined in [17, 18]. A direct substitution of these potentials in equations (2) and (3) instead of the purely phenomenological Gogny or Skyrme interactions results in computationally more expensive numerical integrals that would need to be calculated at every HFB iteration. To sidestep the increase in computational cost two approximations are made. The spatial form factor of the potential in equation (2) is approximated by a sum of Gaussian functions whose strength and range parameters are adjusted to reproduce the original chiral interaction. The use of Gaussian functions allows treating the Hartree term in the same fashion as a Gogny interaction and make use of previously benchmarked many-body DFT codes in which Gogny interactions have already been implemented, such as `HFBTHOv3` [19]. The second approximation is to employ the DME formalism to calculate the Fock energy term. The DME expands the non-local density in equation (3) in terms of a local density and derivatives of that local density as

$$\rho\left(\boldsymbol{R} + \frac{\boldsymbol{r}}{2}, \boldsymbol{R} - \frac{\boldsymbol{r}}{2}\right) \approx \Pi^{\rho}_0(k_F r)\rho(\boldsymbol{R}) + \frac{r^2}{6}\Pi^{\rho}_2(k_F r)\left[\frac{1}{4}\Delta\rho(\boldsymbol{R}) - \tau(\boldsymbol{R}) + \frac{3}{5}k_F^2\rho(\boldsymbol{R})\right], \tag{4}$$

where the Π functions are proportional to a reduced spherical Bessel function, k_F is the usual Fermi momentum and $\tau(\boldsymbol{r})$ is the kinetic density given by

$$\tau(\boldsymbol{r}) = \nabla \cdot \nabla' \rho(\boldsymbol{r}, \boldsymbol{r}')|_{\boldsymbol{r}=\boldsymbol{r}'}. \tag{5}$$

As a result, the DME turns the finite-range part of the local, χ-EFT chiral potential into a Skyrme-like functional, the main difference with traditional Skyrme functionals being that all coupling constants are now functionals of the local density.

The procedure briefly outlined above results in a semi-phenomenological approach in which the re-parametrization of the Skyrme-like part encodes the zero-range, i.e., high-energy, part of the chiral potential and represents the short-range interaction, while the finite-range part of the interaction provides the order-by-order improvement to the description of many-body potentials. With this approach it is in principle possible to construct a scalable framework capable of describing neutron-rich nuclei with a series of systematic improvements.

3. Numerical accuracy of the approximation for the Hartree term

As mentioned in the previous section, the implementation of the microscopically constrained density functionals requires approximating the chiral potential in equation (2) as a sum of Gaussian functions,

$$V_\chi(r) \rightarrow \tilde{V}(r) = \sum_{i=1}^{N} V_i e^{-r^2/\mu_i} \tag{6}$$

It is crucial then to ensure that this Gaussian expansion gives a correct description of the actual chiral interaction in terms of Yukawa potentials.

An important feature of the local chiral potentials of [17, 18] is that the potential contains a regulator in order to avoid divergences at short distances. The regulator has the form

$$f(r) = \left[1 - \exp\left(-\frac{r^2}{R_c^2}\right)\right]^n. \tag{7}$$

The potentials implemented in [12] in particular use $R_c = 1.0$ fm and $n = 6$. This regulator ensures that the interaction vanishes as $r \rightarrow 0$. A simple way to reproduce this kind of behavior in the sum of Gaussian functions in equation (6) is to impose the condition,

$$V_N = -\sum_{i=1}^{N-1} V_i. \tag{8}$$

The choice for the number of Gaussian functions used in the approximation is dictated by the necessary compromise between precision and available resources. Using too few Gaussian functions causes a poor approximation that will introduce numerical errors in the HFB many-body calculation; using too many Gaussian functions significantly increases the necessary memory to store all non-zero one-body density matrix elements with the possibility of going beyond the amount of memory available in current high performance computing architectures. With these restrictions in mind it is worth looking at the difference between the actual chiral interaction and equation (6) as a function of the number of Gaussian terms included in the approximation. While these proceedings focus in particular on the W_C^{NLO} and V_C^{N2LO} terms in the chiral interaction outlined in [17], the analysis presented here applies as well to the other central terms at different orders with and without the inclusion of Δ excitations.

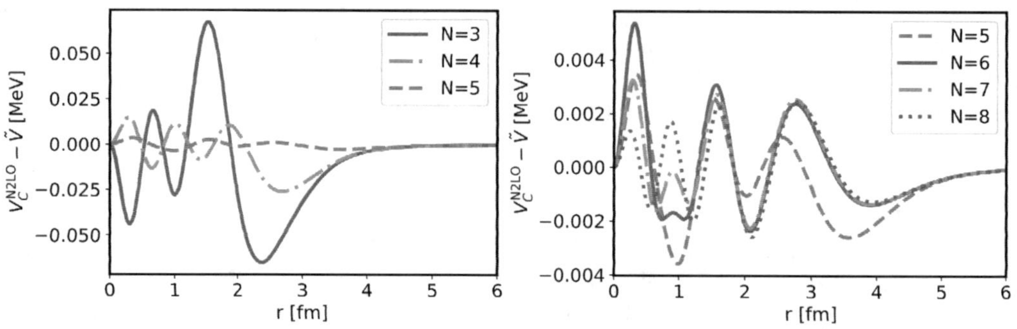

Figure 1. Difference between the regularized central term V_C in the local chiral potential at N2LO as described in [17] and the Gaussian expansion of equation (6) with different number of Gaussian functions N. The $N = 5$ case is included in both panels for comparison

For each of the two interactions mentioned above, we study six different approximations that go from using only three Gaussian functions up to a total of eight. In each approximation the V_i and μ_i parameters are adjusted, imposing the constraint of equation (8), to reproduce the chiral interaction as well as possible. While a regularized approximation ($\tilde{V} \to 0$ as $r \to 0$) can already be constructed using only two Gaussian functions, a reasonable approximation is obtained starting at $N = 3$. The same type of analysis for the central isospin term $W_C^{\rm NLO}$ is presented in figure 2. Although the curves in the right panel show a reduction in the difference between the chiral potential and the sum of Gaussian functions approximation from $N = 5$ to $N = 6$, the order for magnitude remains the same all the way up to $N = 8$. Considering that the computational memory requirements would be considerably increased by an extra Gaussian function it is preferred to keep the approximation at $N = 5$ as the optimal compromise between numerical accuracy and memory requirements.

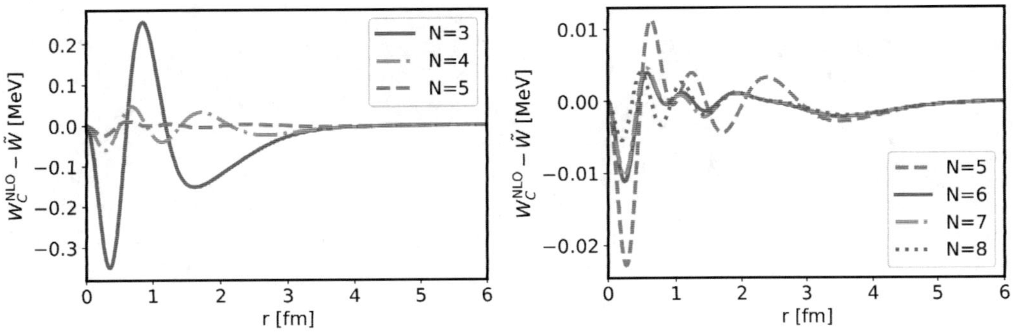

Figure 2. Same as figure 1 for the central isospin-dependent W_C term at NLO

Although the figures 1 and 2 give visual evidence that increasing the number of Gaussian functions beyond $N = 5$ adds little to no improvement, it is preferred to quantify this statement via a numerical calculation. To estimate the average discrepancy between the chiral interaction and the Gaussian approximation (6), we calculate the root mean square deviation using a fine grid sampling in radius every 0.01 fm from $r = 0$ fm to $r = 10$ fm. The numerical results are shown in table 1 and quantifies what was already suggested in the previous figures; there is little to no change in the error beyond $N = 5$.

Table 1. Root mean square deviation between the two central terms V_C and W_C described in [17] and the Gaussian expansion in equation (6) for different number of Gaussian functions.

N	V_c MeV	W_c MeV
3	0.0229	0.0828
4	0.0083	0.0155
5	0.0013	0.0039
6	0.0013	0.0018
7	0.0010	0.0017
8	0.0010	0.0012

4. Conclusions and outlook

The implementation of microscopically-constrained EDF allows performing many-body calculations of large neutron-rich systems with a systematic order-by-order improvement determined by χ-EFT. The performance of this new family of functionals has shown an improved description of nuclear masses when compared to other state-of-the-art functionals like UNEDF2 [12]. This milestone paves the way towards performing calculations relevant to astrophysics, such as simulations of the r-process, with nuclear models with a firm microscopic underpinning stemming from quarks and gluons. Given the potential of these new functionals, it is important to verify that the approximations employed for the implementation have a reasonable level of accuracy in order to avoid compromising the numerical precision of upcoming calculations. These proceedings focused on the numerical accuracy of the sum of Gaussian functions used to represent the central terms of the chiral potential in the calculation of the Hartree energy term. While an infinite number of Gaussian functions could potentially guarantee numerically exact representation of the regularized Yukawa potentials in the chiral interaction, computational resources are limited and a large number of Gaussian functions results in a prohibitively large amount of computational memory. This problem can be mitigated by finding the smallest number of Gaussian functions in the approximation that results in an acceptable level of accuracy.

While these proceedings focused on the sum of Gaussian functions approximation, the numerical accuracy of other approximations and numerical integrals was verified during the implementation of the new family of microscopic functionals. However, other sources of uncertainty remain present and need to be quantified and propagated. For example, beyond mean-field effects near closed shell nuclei, which are not described within the single-reference EDF approach used here, can clearly be seen in figure 5 of [12]. This shows a clear and systematic discrepancy between experimental values and the HFB approach and tools for modeling this discrepancy need to be implemented. An approach based on Bayesian Neural Networks has been proposed for this very particular type of discrepancy in simpler mass models like the liquid drop model [20], and a similar approach can be directly applied to density based calculations within the HFB formalism.

5. Acknowledgments

Support for this work was provided through the Scientific Discovery through Advanced Computing (SciDAC) program funded by U.S. Department of Energy, Office of Science, Advanced Scientific Computing Research and Nuclear Physics. It was partly performed under the auspices of the US Department of Energy by the Lawrence Livermore National Laboratory under Contract DE-AC52-07NA27344. Computing support for this work came from the Lawrence Livermore National Laboratory (LLNL) Institutional Computing Grand Challenge program.

References

[1] Furnstahl R, Serot B D and Tang H B 1997 *Nuclear Physics A* **615** 441–482 ISSN 03759474 URL `http://linkinghub.elsevier.com/retrieve/pii/S0375947496004721`

[2] Bedaque P F, Rupak G, Grießhammer H W and Hammer H W 2003 *Nuclear Physics A* **714** 589–610 ISSN 03759474 URL `http://linkinghub.elsevier.com/retrieve/pii/S0375947402014021`

[3] Ekström A, Baardsen G, Forssén C, Hagen G, Hjorth-Jensen M, Jansen G R, Machleidt R, Nazarewicz W, Papenbrock T, Sarich J and Wild S M 2013 *Physical Review Letters* **110** ISSN 0031-9007, 1079-7114 URL `https://link.aps.org/doi/10.1103/PhysRevLett.110.192502`

[4] Valderrama M P and Phillips D 2015 *Physical Review Letters* **114** ISSN 0031-9007, 1079-7114 URL https://link.aps.org/doi/10.1103/PhysRevLett.114.082502

[5] Hergert H, Yao J M, Morris T D, Parzuchowski N M, Bogner S K and Engel J 2018 *Journal of Physics: Conference Series* **1041** 012007 ISSN 1742-6588, 1742-6596 URL http://stacks.iop.org/1742-6596/1041/i=1/a=012007?key=crossref.a38568cf679a9bdb959aff4d5873fb26

[6] Hagen G, Hjorth-Jensen M, Jansen G R and Papenbrock T 2016 *Physica Scripta* **91** 063006 ISSN 0031-8949, 1402-4896 URL http://stacks.iop.org/1402-4896/91/i=6/a=063006?key=crossref.30c75e8d4e4045fe105b7011a01925f6

[7] Engel J, Bender M, Dobaczewski J, Nazarewicz W and Surman R 1999 *Physical Review C* **60** ISSN 0556-2813, 1089-490X URL https://link.aps.org/doi/10.1103/PhysRevC.60.014302

[8] Surman R, McLaughlin G C and Hix W R 2006 *The Astrophysical Journal* **643** 1057–1064 ISSN 0004-637X, 1538-4357 URL http://stacks.iop.org/0004-637X/643/i=2/a=1057

[9] Schunck N 2019 *Energy density functional methods for atomic nuclei.* IOP Expanding Physics (Bristol, UK: IOP Publishing) URL https://iopscience.iop.org/book/978-0-7503-1422-0

[10] Bender M, Heenen P H and Reinhard P G 2003 *Rev. Mod. Phys.* **75** 121 URL http://rmp.aps.org/abstract/RMP/v75/i1/p121_1

[11] Stoitsov M, Kortelainen M, Bogner S K, Duguet T, Furnstahl R J, Gebremariam B and Schunck N 2010 *Phys. Rev. C* **82** 054307 URL https://link.aps.org/doi/10.1103/PhysRevC.82.054307

[12] Navarro Pérez R, Schunck N, Dyhdalo A, Furnstahl R J and Bogner S K 2018 *Physical Review C* **97** ISSN 2469-9985, 2469-9993 URL https://link.aps.org/doi/10.1103/PhysRevC.97.054304

[13] Dyhdalo A, Bogner S K and Furnstahl R J 2017 *Physical Review C* **95** ISSN 2469-9985, 2469-9993 URL http://link.aps.org/doi/10.1103/PhysRevC.95.054314

[14] Skyrme T H R 1959 *Nucl. Phys.* **9** 615 URL https://doi.org/10.1016/0029-5582(58)90345-6

[15] Stone J and Reinhard P G 2007 *Prog. Part. Nucl. Phys.* **58** 587 URL https://doi.org/10.1016/j.ppnp.2006.07.001

[16] Dechargé J and Gogny D 1980 *Phys. Rev. C* **21** 1568 URL https://doi.org/10.1103/PhysRevC.21.1568

[17] Piarulli M, Girlanda L, Schiavilla R, Navarro Pérez R, Amaro J E and Ruiz Arriola E 2015 *Physical Review C* **91** 024003 URL https://link.aps.org/doi/10.1103/PhysRevC.91.024003

[18] Piarulli M, Girlanda L, Schiavilla R, Kievsky A, Lovato A, Marcucci L E, Pieper S C, Viviani M and Wiringa R B 2016 *Physical Review C* **94** 054007 URL https://link.aps.org/doi/10.1103/PhysRevC.94.054007

[19] Navarro Pérez R, Schunck N, Lasseri R D, Zhang C and Sarich J 2017 *Computer Physics Communications* **220** 363–375 ISSN 00104655 URL https://linkinghub.elsevier.com/retrieve/pii/S0010465517302047

[20] Utama R, Piekarewicz J and Prosper H B 2016 *Physical Review C* **93** ISSN 2469-9985, 2469-9993 URL https://link.aps.org/doi/10.1103/PhysRevC.93.014311

IOP Conf. Series: Journal of Physics: Conf. Series **1308** (2019) 012015 doi:10.1088/1742-6596/1308/1/012015

Electromagnetic couplings of pentaquarks

E. Ortiz-Pacheco and R. Bijker

Instituto de Ciencias Nucleares, Universidad Nacional Autónoma de México, A.P. 70-543, 04510 Ciudad de México, México

E-mail: bijker@nucleares.unam.mx

Abstract. In this contribution, we discuss the electromagnetic couplings of pentaquark states with hidden charm. This work is motivated by recent experiments at CERN by the LHCb Collaboraton and current experiments at JLab to confirm the existence of hidden-charm pentaquarks in photoproduction experiments.

1. Introduction

The observation of hidden-charm pentaquark states by the LHCb Collaboration [1]-[5] has created an avalanche of theoretical studies on the nature of these states and on different plausible interpretations of the observed signals, *e.g.* kinematical effects [6, 7, 8], molecular states [9]-[15] and compact pentaquarks [16]-[23], meson-baryon states coupled to a five-quark compact core [24] and a baryo-quarkonium states [25]. The existence of narrow N^* and Λ^* resonances with hidden charm was predicted [26, 27] in a coupled-channel unitary approach five years before the LHCb data were published, as well as in Refs. [28]. More information on the experimental and theoretical aspects of pentaquark states, as well as a more complete list of references, can be found in the reviews [29]-[34].

In the present contribution we discuss the electromagnetic couplings of $uudc\bar{c}$ hidden-charm pentaquark states which are relevant for photoproduction experiments at JLab [35]-[40].

2. Pentaquark states

Pentaquark states depend both on the orbital degrees of freedom and the internal degrees of freedom of color, spin and flavor

$$\psi = \psi^{\rm o}\psi^{\rm c}\phi^{\rm f}\chi^{\rm s} \; . \tag{1}$$

The construction of the classification scheme of $uudc\bar{c}$ pentaquark states was carried out explicitly in Ref. [23, 41] using the following two conditions: (i) the pentaquark wave function should be a color singlet and (ii) the wave function of the four-quark subsystem should be antisymmetric. The permutation symmetry of four-quark states is characterized by the S_4 Young tableaux [4], [31], [22], [211] and [1111] or, equivalently, by the irreducible representations of the tetrahedral group $\mathcal{T}_d$ (which is isomorphic to S_4) as A_1, F_2, E, F_1 and A_2, respectively.

The first condition that the pentaquark wave function has to be a color-singlet, implies that the color wave function of the four-quark configuration has to be a [211] triplet with F_1 symmetry under $\mathcal{T}_d$. As a consequence, the second condition that the total q^4 wave function

IOP Conf. Series: Journal of Physics: Conf. Series **1308** (2019) 012015 doi:10.1088/1742-6596/1308/1/012015

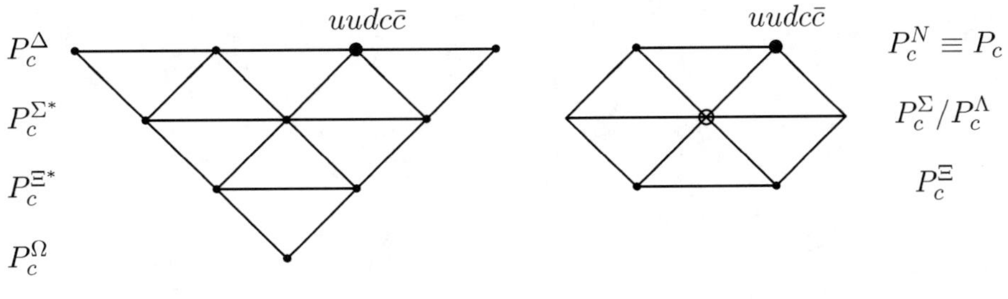

Figure 1. Pentaquark decuplet and octet

has to be antisymmetric (A_2), means that the orbital-spin-flavor part is a [31] triplet with F_2 symmetry

$$\psi = \left[\psi_{F_1}^{\rm c} \times \psi_{F_2}^{\rm osf}\right]_{A_2} , \tag{2}$$

where the subindices refer to the symmetry properties of the four-quark subsystem under permutation. Moreover, in this contribution we limit ourselves to ground-state pentaquark states, *i.e.* without orbital excitations, which are symmetric (A_1). Therefore, the spin-flavor part is a [31] state with F_2 symmetry

$$\psi = \psi_{A_1}^{\rm o} \left[\psi_{F_1}^{\rm c} \times \psi_{F_2}^{\rm sf}\right]_{A_2} . \tag{3}$$

In Ref. [23] it was shown that there are in total seven $uudc\bar{c}$ ground-state pentaquark configurations with angular momentum and parity $J^P = 3/2^-$ (which is quoted in the literature as the most likely value of the angular momentum and parity of the P_c pentaquark [1]), three of which belong to a flavor decuplet and the remaining four to a flavor octet (see Fig. 1 and first column of Table 1).

3. Electromagnetic couplings

For experiments that aim to study pentaquarks through near threshold J/ψ photoproduction at JLab, the size of the electromagnetic couplings of the pentaquarks is important. Here we discuss the electromagnetic couplings for the ground state pentaquarks with spin and parity $J^P = 3/2^-$. Electromagnetic couplings are described by

$$\mathcal{H}_{\rm em} = e \int d^3x \, \hat{J}^{\mu}(\vec{x}) A_{\mu}(\vec{x}) , \tag{4}$$

where J^{μ} is the electromagnetic current summed over all quark flavors

$$\hat{J}^{\mu}(\vec{x}) = \sum_q e_q \, \bar{q}(\vec{x}) \gamma^{\mu} q(\vec{x}) . \tag{5}$$

The electromagnetic coupling of Eq. (4) describes both the emission (and absorption) of the photon off a quark, as is used in studies of photocouplings of baryons [42], and the annihilation process of a quark-antiquark pair [43]. For the process of interest, $P_c \to N + \gamma$, the relevant term is the annilation of a pair of $c\bar{c}$ quarks, Fig. 2. In the present calculation we use the nonrelativistic form of the interaction. The radiative decay widths can be calculated as [44]

$$\Gamma(P_c \to N + \gamma) = \frac{\rho}{(2\pi)^2} \frac{2}{2J+1} \sum_{\nu>0} |A_{\nu}(k)|^2 , \tag{6}$$

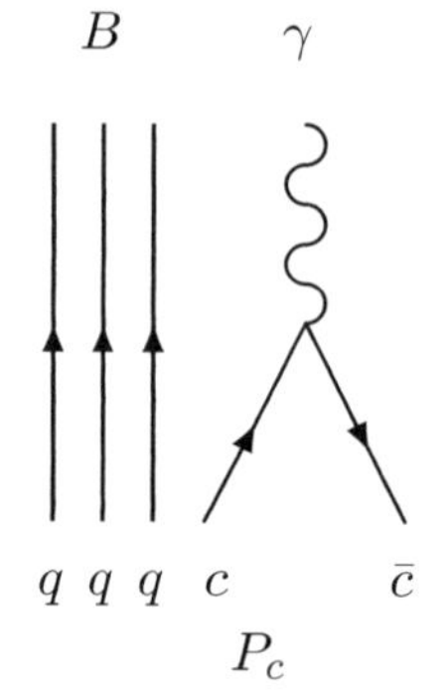

Figure 2. Electromagnetic decay of pentaquark P_c into a baryon B and a photon, $P_c \to B+\gamma$.

where ρ is the phase space factor, and A_ν denotes the helicity amplitude

$$A_\nu(k) \;=\; \langle N, 1/2^+, \nu-1; \gamma\, |\mathcal{H}^{\rm nr}_{\rm em}|\, P_c, 3/2^-, \nu\rangle \;=\; \sqrt{\frac{4\pi\alpha}{k_0}}\,\beta_\nu\, F(k)\;. \tag{7}$$

Here α is the fine-structure constant, k_0 and $k = |\vec{k}|$ represent the energy and the momentum of the photon. The coefficient β_ν is the contribution from the color-spin-flavor part for the annihilation of a $c\bar{c}$ color-singlet pair with spin $S = S_z = 1$. It is straightforward to show that for the cases considered, *i.e.* ground-state pentaquarks with $J^P = 3/2^-$, the helicity amplitudes are related by

$$\beta_{1/2} \;=\; \beta_{3/2}/\sqrt{3}\;. \tag{8}$$

Finally, $F(k)$ is a form factor denoting the contribution from the orbital part of the pentaquark wave function. Its specific form depends on the type of quark model used: harmonic oscillator, hypercentral, or other. Here we concentrate on the color-spin-flavor part which is common to all quark models. In Table 1 we show the results for the contribution from the color-spin-flavor part to the helicity amplitudes for different configurations of $uudc\bar{c}$ pentaquarks.

Table 1. Contribution from the color-spin-flavor part to the helicity amplitudes for the electromagnetic decays of $uudc\bar{c}$ decuplet (top) and octet (bottom) pentaquark states into $N+\gamma$. Here e_c is the electric charge of the charm quark $e_c = 2/3$.

State	Name	$\beta_{1/2}$	$\beta_{3/2}$
$[\phi_{A_1} \times \chi_{F_2}]_{F_2}$	P_c^Δ	0	0
$[\phi_{F_2} \times \chi_{A_1}]_{F_2}$	P_c^Δ	0	0
$[\phi_{F_2} \times \chi_{F_2}]_{F_2}$	P_c^Δ	0	0
$[\phi_{F_2} \times \chi_{A_1}]_{F_2}$	P_c^N	0	0
$[\phi_{F_2} \times \chi_{F_2}]_{F_2}$	P_c^N	$\frac{1}{6\sqrt{2}}e_c$	$\frac{1}{2\sqrt{6}}e_c$
$[\phi_E \times \chi_{F_2}]_{F_2}$	P_c^N	$-\frac{1}{6}e_c$	$-\frac{1}{2\sqrt{3}}e_c$
$[\phi_{F_1} \times \chi_{F_2}]_{F_2}$	P_c^N	$-\frac{1}{2\sqrt{6}}e_c$	$-\frac{1}{2\sqrt{2}}e_c$

The couplings to the octet configuration with ϕ_{A_1} and the three decuplet configurations vanish because of symmetry reasons. The strongest coupling is to the octet pentaquark configuration with ϕ_{F_1}, followed by ϕ_E and ϕ_{F_2}.

4. Summary and conclusions

In conclusion, in this contribution we discussed the electromagnetic couplings of ground-state $uudc\bar{c}$ pentaquark states with angular momentum and parity $J^P = 3/2^-$. At present we did not include orbital excitations. Of the seven possible configurations only three octet configurations have a nonvanishing photocoupling. Since the photon momentum is large, we expect that these couplings are strongly suppressed by the form factor, $F(k)$, which represents the contribution from the orbital part of the pentaquark wave function.

If the signal observed by the LHCb Collaboration indeed corresponds to hidden-charm pentaquarks, there should be an entire multiplet of pentaquark states, for example a pentaquark octet consisting of P_c^N, P_c^Σ, P_c^Λ and P_c^Ξ states.

Acknowledgments

This work was supported in part by grant IN109017 from DGAPA-UNAM, Mexico and grants 251817 and 340629 from CONACyT, Mexico.

References

[1] Aaij R *et al.* (LHCb Collaboration) 2015 *Phys. Rev. Lett.* **115** 072001
[2] Aaij R *et al.* (LHCb Collaboration) 2016 *Phys. Rev. Lett.* **117** 082002
[3] Aaij R *et al.* (LHCb Collaboration) 2016 *Phys. Rev. Lett.* **117** 082003
[4] Aaij R *et al.* (LHCb Collaboration) 2017 *Phys. Rev. Lett.* **118** 022003
[5] Aaij R *et al.* (LHCb Collaboration) 2019 *Phys. Rev. Lett.* in press [arXiv:1904.03947]
[6] Guo F K, Meissner U G, Wang W and Yang Z 2015 *Phys. Rev. D* **92** 071502
[7] Liu X H, Wang Q and Zhao Q 2016 *Phys. Lett. B* **757** 231
[8] Mikhasenko M 2015 arXiv:1507.06552
[9] Karliner M and Rosner J L 2015 *Phys. Rev. Lett.* **115** 122001
[10] Chen R, Liu X, Li X Q and Zhu S L 2015 *Phys. Rev. Lett.* **115** 132002
[11] Chen H X, Chen W, Liu X, Steele T G and Zhu S L 2015 *Phys. Rev. Lett.* **115** 172001
[12] Roca L, Nieves J and Oset E 2015 *Phys. Rev. D* **92** 094003
[13] He J 2016 *Phys. Lett. B* **753** 547
[14] Eides M I, Petrov V Yu, Polyakov M V 2016 *Phys. Rev. D* **93** 054039
[15] Yamaguchi Y and Santopinto E 2017 *Phys. Rev. D* **96** 014018
[16] Maiani L, Polosa A D and Riquer V 2015 *Phys. Lett. B* **749** 289
[17] Lebed R F 2015 *Phys. Lett. B* **749** 454
[18] Wang G J, Chen R, Ma L, Liu X and Zhu S L 2016 *Phys. Rev. D* **94** 094018
[19] Yang G, Ping J and Wang F 2017 **Phys. Rev. D 95** 014010
[20] Deng C, Ping J, Huang H and Wang F 2017 *Phys. Rev. D* **95** 014031
[21] Takeuchi S and Takizawa M 2017 *Phys. Lett. B* **764** 254
[22] Santopinto E and Giachino A 2017 *Phys. Rev. D* **96** 014014
[23] Ortiz-Pacheco E, Bijker R and Fernández-Ramírez C 2019 *J. Phys. G: Nucl. Part. Phys.* **46** 065104 [arXiv:1808.10512]
[24] Yamaguchi Y, Giachino A, Hosaka A, Santopinto E, Takeuchi S and Takizawa M 2017 *Phys. Rev. D* **96** 114031
[25] Ferretti J, Santopinto E, Naeem Anwar M and Bedolla M A 2019 *Phys. Lett. B* **789** 562
[26] Wu J J, Molina R, Oset E and Zou B S 2010 *Phys. Rev. Lett.* **105** 232001
[27] Wu J J, Molina R, Oset E and Zou B S 2011 *Phys. Rev. C* **84** 015202
[28] Yang Z C, Sun Z F, He J, Liu X and Zhu S L 2012 *Chin. Phys. C* **36** 6
[29] Chen H X, Chen W, Liu X and Zhu S L 2016 *Phys. Rep.* **639** 1
[30] Esposito A, Pilloni A and Polosa A D 2016 *Phys. Rep.* **668** 1
[31] Ali A, Lange J S and Stone S 2018 *Prog. Part. Nucl. Phys.* **97** 123
[32] Olsen S L, Skwarnicki T and Zieminska D 2018 *Rev. Mod. Phys.* **90** 015003
[33] Karliner M, Rosner J L and Skwarnicki T 2018 *Ann. Rev. Nucl. Part. Sci.* **68** 17

[34] Guo F-K, Hanhart C, Meiner U-G, Wang Q, Zhao Q and Zou B-S 2018 *Rev. Mod. Phys.* **90** 015004
[35] Kubarovsky V and Voloshin M B 2015 *Phys. Rev. D* **92** 031502(R)
[36] Wang Q, Liu X H and Zhao Q 2015 *Phys. Rev. D* **92** 034022
[37] Karliner M and Rosner J L 2016 *Phys. Lett. B* **752** 329
[38] Hiller Blin A N, Fernández-Ramírez C, Jackura A, Mathieu V, Mokeev V I, Pilloni A and Szczepaniak A P 2016 *Phys. Rev. D* **94** 034002
[39] Fernández-Ramírez C, Hiller Blin A N and Pilloni A 2017 *arXiv:1703.06928*
[40] Meziani Z E *et al.* 2016 *arXiv:1609.00676*
[41] Bijker R 2017 *J Phys Conf Ser* **876** 012004
[42] Copley L A, Karl G and Obryk E 1969 *Nucl Phys B* **13** 303
[43] Le Yaouanc A, Oliver L, Pene O and Raynal J 1988 *Hadron Transitions in the Quark Model* Gordon and Breach
[44] Bijker R, Iachello F and Leviatan A 2000 *Ann. Phys. (N.Y.)* **284** 89 (2000)

IOP Conf. Series: Journal of Physics: Conf. Series **1308** (2019) 012016 doi:10.1088/1742-6596/1308/1/012016

^{19}F(p,α)^{16}O and ^{19}F(α,p)^{22}Ne Reaction Rate Measured via THM and Fluorine Nucleosynthesis in AGB stars

S. Palmerini[1,2], G. D'Agata[3], M. La Cognata[4], I. Indelicato[4], R. G. Pizzone[4], O. Trippella[2] and D. Vescovi[2,5]

[1] Dipartimento di Fisica e Geologia, Università degli Studi di Perugia, via A. Pascoli, I-06125 Perugia, Italy
[2] I.N.F.N. sezione di Perugia, via A. Pascoli, I-06125 Perugia, Italy
[3] Nuclear Physics Institute of the Czech Academy of Sciences, p.r.i., Řež 130, Řež, 25068 Czech Republic
[4] I.N.F.N. Laboratori Nazionali del Sud, Via Santa Sofia 62, I-95123, Catania, Italy
[5] Gran Sasso Science Institute, Viale Francesco Crispi, 7, 67100 L'Aquila, Italy

E-mail: `sara.palmerini@pg.infn.it`

Abstract. Asymptotic Giant Branch (AGB) stars have been proven to be sites of F production through spectroscopy observations by several authors, but it is not clear whether these stars might account for the total fluorine abundance of the Galaxy. Recently the two main channels for ^{19}F destruction in AGB stars, namely the ^{19}F(α,p)^{22}Ne and ^{19}F(p,α)^{16}O reactions, have been studied via the Trojan Horse Method in the energy range of interest for astrophysics. In both cases experimental results have shown the presence of resonant structures below 500 keV, hinting to an enhancements in efficiency of fluorine destruction by stellar H- and He- burning. In particular the ^{19}F(p,α)^{16}O reaction rate at $T_9 \leq 0.2K$ turns out to be increased up to a factor of 1.7 while the ^{19}F(p,α)^{16}O is enhanced more than a factor of 4 at $0.1 \leq T_9 \leq 0.25$. We present here a re-analysis of the role of AGB stars as fluorine galactic source by comparing stellar observations with predictions of AGB nucleosynthesis (for stellar masses from 1.5 to 5M$_\odot$) computed by employing in state-of-the-art models the THM reaction rates for ^{19}F destruction.

1. Introduction

The sole stable isotope of fluorine, the ^{19}F, is produced in stars trough a very complicate network of reactions and it can be easily destroyed by both proton- and α-captures [1]. Since all stars experience phases of H-burning and He-burnig during their evolutions the sites that could host the nucleosynthesis of fluorine are many. In particular, four sites have been proposed.

1. Asymptotic Giant Branch (hereafter AGB) stars are suggested to produce fluorine by means of two different processes. At solar metallicity the main chain of F production starts with ^{14}N and includes α, neutron and proton captures. While, at lower metallicities, the fluorine production is suggested to be dependent on ^{13}C ([2] and references therein).
2. Type II supernovae (SNe II) are suggested to produce cosmic fluorine via the neutrino process. Indeed the enormous quantity of neutrinos released during the core collapse could convert a significant amount of ^{20}Ne into fluorine [3].
3. Wolf-Rayet (W-R) stars, which produce F via a nucleosynthesis network similar to the one burned by AGB stars, might enrich in fluorine the interstellar medium thanks to their strong stellar winds [4, 5].
4. Hydrogen-deficient stars formed by white-dwarf merging were shown to have high carbon and fluorine abundances by [6]. However, the number of stars of this type is so small compared to the AGB stars, SNe II, and W-R stars, that their contribution to the galactic fluorine abundance is negligible.

IOP Conf. Series: Journal of Physics: Conf. Series **1308** (2019) 012016 doi:10.1088/1742-6596/1308/1/012016

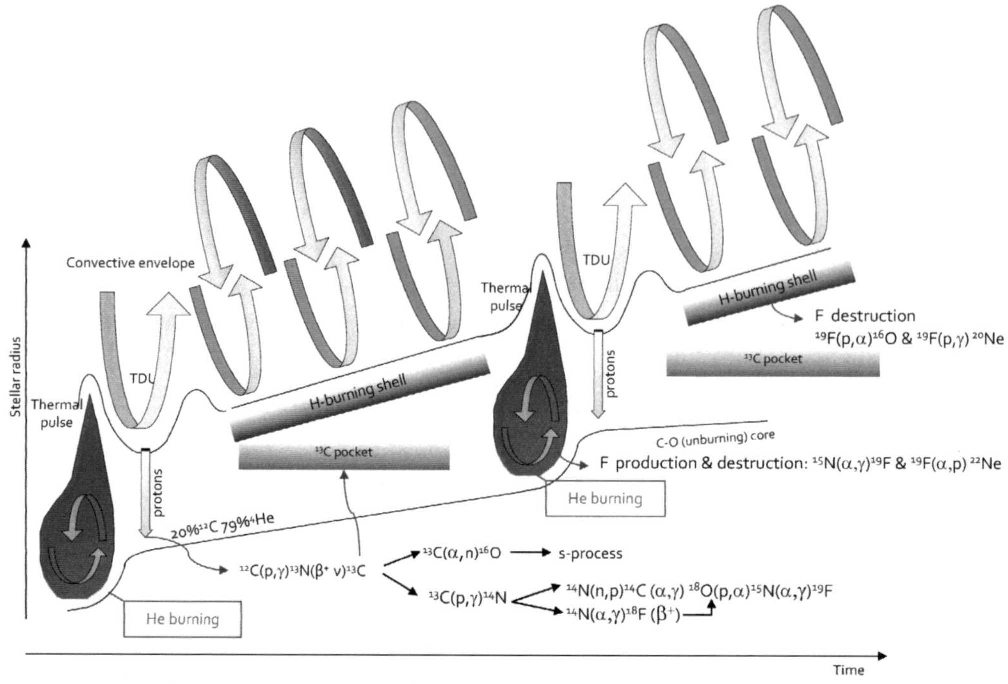

Figure 1. Cartoon showing the temporal evolution of the internal structure of an AGB star. The "drop" structures are thermal pulses (see text for details) and the sites of fluorine production and destruction are indicated by arrows linked to the nucleosynthesis reactions that take place in the different regions of the star. The labels indicate the places of the H-burning shell, the He-shell, the ^{13}C-pocket and the convective envelope.

AGB stars are commonly considered sites for the production of fluorine, because of the enhanced abundance of the element observed in a large sample of stars showing the characteristic line at $\lambda = 2.3358\ \mu$m in their spectra (see e.g. [7] and references therein). However the the question is not whether AGB stars synthesize fluorine, but whether their production is sufficient to account for its cosmic abundance. A way to determine the site of fluorine production and to disentangle among the nucleosynthesis mechanisms suggested is to compare the observed fluorine abundance with those of other elements, whose stellar origins are well known, e.g, oxygen that is produced mainly in SNe II. This was done by [8], which observed [F/Fe] versus [Fe/H] and [F/O] versus [O/H] trends in a sample of 49 nearby, bright K giants. Both the trends were found to increase and this fact hints that ν-process could not be the dominant fluorine producer in the solar neighborhood. This findings foreseen the hypothesis that AGB stars are the main site of fluorine production in galaxies even if other authors, using different models of galactic chemical evolution, state that the yields from AGB stars are not enough to account for the whole fluorine abundance in the Milky Way [7].

In principle the resulting abundance of a nucleus (or an element) can be larger than theoretical predictions because the production channels are more efficient than thought or because destruction mechanisms are less. In this paper we investigate this latter hypothesis by employing in a code for AGB nucleosynthesis calculations the ^{19}F$(\alpha,p)^{22}$Ne and ^{19}F$(p,\alpha)^{16}$O reaction rate recently measured at astrophysical energies by the Trojan Horse Method (hereafter THM).

IOP Conf. Series: Journal of Physics: Conf. Series **1308** (2019) 012016 doi:10.1088/1742-6596/1308/1/012016

2. Fluorine Nucleosynthesis in AGB stars

The AGB phase represents an evolved stage of stars with mass smaller than 6-8$M_{\odot}$. During this phase the stellar core (made of C and O) is degenerate and the main source of energy is the thin radiative H-burning shell placed below the extended and cold convective envelope. Between the H-shell and the CO core there is a region rich in He (the He-shell), where episodes of convective He-burning periodically occur, these instabilities are known as *thermal pulses* (TP). At the end of each pulse the convective envelope penetrates into the inner stellar region bringing into the surface the fresh product of the stellar nucleosynthesis, this phenomenon is known as *third dredge-up* (hereafter, TDU).

AGB stars have been extensively studied in the last 30 years because they are important sites for the production of nuclei heavier than Fe via the s-processing (or slow neutron captures). It takes place in the He-shell during the TPs, thanks to neutrons delivered by the ^{22}Ne(α,n)^{25}Mg reaction, as well as during the H-burning phase in the He-rich intershell region where neutrons are released by the ^{13}C(α,n)^{16}O reaction. The formation of a ^{13}C reservoir just below the H-shell has been investigated by several authors because of its crucial role for the s-process. In this paper we will not enter in details of the so called ^{13}C-pocket formation, but it is important to remind that this is due to an injection of protons in the stellar layers rich in ^{4}He and ^{12}C during the TDU, which triggers the reaction chain: ^{12}C(p,γ)^{13}N(β)^{13}C. Whenever the amount of injected protons is sufficiently large, also ^{14}N is produced by further proton captures on ^{13}C. Since ^{14}N has a very large neutron capture cross section, its production has to be avoided to guarantee an efficient production of heavy nuclei via the s-process. However, the ^{14}N is the seed of the ^{19}F production, which follows the path ^{14}N(α,γ)^{18}F($\beta^{+}\nu$)^{18}O(p,α)^{15}N(α,γ)^{19}F or, alternatively, the chain ^{18}O(p,α)^{15}N(α,γ)^{19}F, if protons are released by the ^{14}N(n, p)^{14}C reaction [9, 10]. When the He-rich materials are engulfed in a TP (at T $\geq 2.5{\cdot}10^{8}K$) part of the ^{19}F is consumed by the ^{19}F(α,p)^{22}Ne , which is the main channel for its consumption in low-mass AGB stars. In this contest if some unburned ^{13}C is still available a further contribution to ^{19}F might still be possible via the ^{15}N(α,γ)^{19}F reaction, but this rare possibility will not be considered in this work.

A further destruction of the F can also be operated by the H-burning shell, where the temperatures are not high enough to produce ^{19}F via proton capture on ^{16}O, but are in any case sufficient for its burning by the ^{19}F(p,α)^{16}O and the ^{19}F(p,γ)^{20}Ne reaction (being the first the most efficient one). Furthermore, proton captures at a few $10^{7}K$ coupled with non convective exchanging of matter between the border of the convective envelope and the H-burning shell (called extra-mixing [21]) can reduce the surface abundance of fluorine in AGBs of less than 3 $M_{\odot}$ [11].

A scheme of the complicate network of reactions responsible for fluorine nucleosynthesis is shown in Figure 1 as long with the temporal evolution of the internal structure of an AGB star.

3. Investigation of the ^{19}F(p,α)^{16}O and ^{19}F(α,p)^{22}Ne reaction via the Trojan Horse Method

Despite its importance, before 2011 [12], no data were available for the ^{19}F(p,α)^{16}O at astrophysical energies ($E_{c.m.} \leq$ 300 keV, where fluorine burning is most effective and widely adopted compilations such as the NACRE [13] used data from different sources to supply a recommended astrophysical factor and reaction rate. Firstly [12] observed three resonances in the energy regions of interest for the study of stellar nucleosynthesis by using the indirect experimental technique called Trojan Horse Method. According with [14], and references therein, through this technique a two body reaction (having two massive particle in the exit channel) can be measured down to the Gamow-peak energy region by performing a three-body reaction at a beam energy that allows to overcome the Coulomb barrier in the reaction entrance channel as well as the electron screening. Via the THM the cross-section of the two-body ^{19}F(p,α)^{16}O has been determined by properly selecting the quasi-free contribution of the three-body reaction ^{2}H(^{19}F, α^{16}O)n, where n plays the role of the spectator particle and ^{2}H acts as the TH nucleus because of its p$\oplus$n cluster structure.

Three THM experiments were performed [12, 11, 15] that have shown the presence of resonant structures at $E_{cm} \leq$ 0.6MeV and hinted to an enhancement of he reaction rate at astrophysical temperatures (about $10^{7} - 10^{8}K$). In particular around the E_{cm} 0.794 MeV resonance, all the data agree reasonably well, while in the E_{cm}0.6MeV energy range, the S(E) factors extracted from the most recent experiment [15] is larger because of a resonance at 251 keV, which had not been observed before by [12, 11], but whose presence was also confirmed by

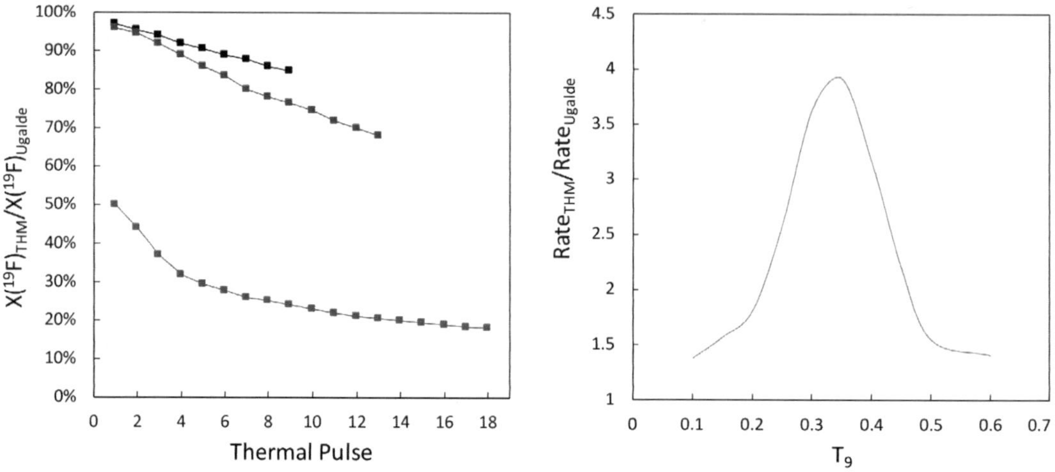

Figure 2. Left panel. Temporal evolution of the ratio between the ^{19}F abundance in the He-shell obtained by adopting the ^{19}F(α,p)^{22}Ne reaction rate determined via THM and the one calculated by [17]. Each dot refers to a thermal pulse (see text for details). The black, the blue and the red curve refer to AGB models of 1.5, 3, and 5 $M_\odot$, respectively. Right panel. Ratio between the THM recommended value of the ^{19}F(α,p)^{22}Ne rate and the value recommended by [17], as a function of the temperature in units of GK.

the data extracted from [16]. From R-matrix calculation it turns out that the difference among the THM reaction rate at T_9=0.04-0.2 are small ($\sim$ 10%) while larger discrepancy ($\sim$ 30%) is found at a temperature T_9 $\geq$0.04. For a comparison of the reaction rates and details on data analysis we refer to figures 13-16 in [15].

The THM has been successfully applied also to investigate the ^{19}F(α,p)^{22}Ne, which is hard to be directly studied because the Coulomb barrier is at about 3.1MeV and the energy range typical of stellar He-burning is 200-1100keV. In this framework the lowest energy at which the reaction has been directly measured is 660 keV [17] and data available in literature were affected by large uncertainties. The reaction chosen to investigate the ^{19}F(α,p)^{22}Ne via the THM was the ^{19}F(^{6}Li, p^{22}Ne)^{2}H because of the well-known cluster structure of the ^{6}Li nucleus ($\alpha \oplus d$), which under proper kinematical conditions allows us to consider the deuteron particle as a spectator. The common findings of [18] and [19] is an increase of the ^{19}F(α,p)^{22}Ne reaction rate up to a factor of 4 respect to the direct data available in literature [17] in the energy region of stellar nucleosynthesis. Such an enhancement can be seen in Table 4 of [19] and in the right panel of figure 2.

4. THM Reaction Rate Implications in AGB star nucleosynthesis

Our study of the implications of the THM reaction rates of the ^{19}F(p,α)^{16}O and ^{19}F(α,p)^{22}Ne to fluorine nucleosynthesis in AGB stars was performed in two steps. Firstly the fluorine production/destruction in the He-rich stellar layers (during both the H-burning periods and the TPs) was investigated in the light of the ^{19}F(α,p)^{22}Ne rate estimated by the THM [18, 19]. In particular, calculations for three stellar models of 1.5, 3, and 5 $M_\odot$ and solar metallicity were performed by the NEWTON code [20]. To better appreciate the effects of the α-capture reaction rate, phenomena of ^{19}F destruction due to proton captures, such as the extra-mixxing ([21] and references therein) or hot bottom burning (in the case of the 5$M_\odot$ model, [22]) were not considered in the first step of our analysis. As explained before, the mechanism for formation of the ^{13}C-pocket might have important implication also for the fluorine nucleosynthesis, in this paper we adopt the profile for proton injection in the inter-shell and the resulting pockets of ^{13}C and ^{14}N suggested by [23] and [24], the same cross-sections for neutron capture reactions used by the quoted authors, and the reaction rates of proton and alpha captures reported in Table 5 by [19].

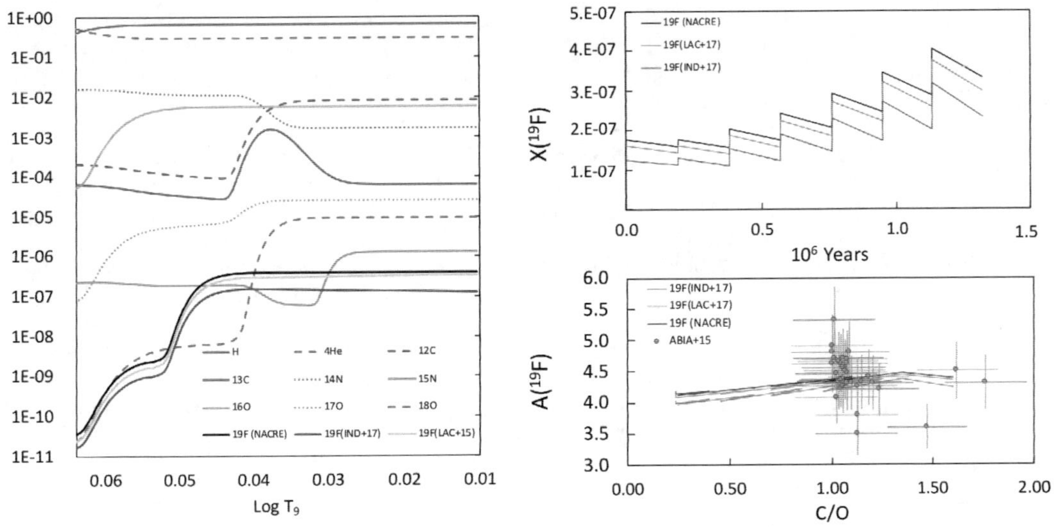

Figure 3. Left panel. Isotopic composition of the H-shell of our $2M_\odot$ and solar metallicity AGB model. Abundances of F and of the nuclei participating into the CNO cycle are shown as a function of the logarithm of the temperature measured in GK. The stellar radius proceeds from the right to the left. Three curves deal with the fluorine abundance profile computed by using the ^{19}F(p,α)^{16}O reaction rate by [13] (black line), [11] (green line) and [15] (red line). Upper right panel. Temporal evolution of the surface abundance of ^{19}F in the $2M_\odot$ and solar metallicity AGB model. The stepwise trend is due to the enrichment in F at each TDU (which brings into the envelope the ashes of the He-burning) and the F destruction of the extra-mixing during the H-burning periods. The three curves show the model outputs obtained with the three ^{19}F(p,α)^{16}O reaction rate we are studying. Lower right panel. Comparison between the same model output reported in the upper panel and the abundances observed in a sample of AGBs by [7]. The F abundance are given using the definition A(^{19}F) = 12 + log(X(^{19}F)/H), being (X^{19}F/H) the abundance of ^{19}F by number, and are reported as a function of the C/O ratio.

The left panel of Figure 2 shows the temporal evolution during the whole AGB phase of the ratio between the ^{19}F abundance obtained in the He-rich stellar region after each TP by adopting the ^{19}F(α,p)^{22}Ne reaction rate determined via THM and the one calculated by using the rate published in [17]. The black, the blue and the red curve refer to AGB models of 1.5, 3, and 5 $M_\odot$, respectively. As illustrated by the right panel of figure 2, at the typical temperature of the He-shell burning, the THM rate is always larger than the one by [17]. As a consequence, ^{19}F is more easily destroyed during TP and the abundances predicted in the stellar interior are smaller. The 5 $M_\odot$ AGB model shows the largest sensitivity to the reaction rate used in calculations, indeed in its He-shell the ^{19}F abundance is reduced down to a factor of 4 in the last pulses. This is so because the He-burning temperature reaches $3.6{\cdot}10^8 K$ that corresponds to the energy at which the difference between the THM cross-section and the [17] one is maximum. Smaller variations are registered in the 1.5 and 3 $M_\odot$ models, where temperatures in the He-shell are lower. However, once the ashes of the He-burning are brought into stellar surface by TDU, the nucleosynthesis products are diluted with envelope materials and the effects of the ^{19}F(α,p)^{22}Ne reaction rate become negligible. None of the studied cases shows a variation in the ratio larger than 5%. An extended discussion on the implication to the F production/destruction in AGB stars of the ^{19}F(α,p)^{22}Ne rate measured by the THM can be found in [19], these authors also present the effects of using the upper limit and the lower limit of the rate in nucleosynthesis calculations.

In the second phase of our analysis the destruction of fluorine due to burning of the ^{19}F(p,α)^{16}O reaction in the H-shell was studied. In doing so the composition of the H-burning

IOP Conf. Series: Journal of Physics: Conf. Series **1308** (2019) 012016 doi:10.1088/1742-6596/1308/1/012016

shell of $2M_\odot$ and solar metallicity AGB stars was computed by the SHELL nucleosynthesis code [21, 25]. The left panel of figure 3 shows the three profile of the ^{19}F abundance computed by using the ^{19}F+p rata published by [11], [15] and [13], being the first two measured by the THM. As expected the fluorine abundances are lower in models that employ a larger reaction rate. Moreover, the post-process code MAGIC [21, 25] was run to estimate the dilution of the surface F abundance due to extra-mixing[1] in the light of the three rates of the ^{19}F(p,α)^{16}O reaction ([11], [15] and [13]). Results are reported in the upper panel on the right of Figure 3, where the profiles of the fluorine abundance in the envelope of a $2M_\odot$ and solar metallicity AGB are drawn as a function of the time, from the beginning of the AGB phase. Finally, in the lower panel on the right of Figure 3, the predictions of our models are compared with the F abundance observed in a sample of AGB stars [7]. Despite the choice made for the ^{19}F(p,α)^{16} cross section, the three models are all in a quite good agreement with the observations. All reported results refer to calculations and comparisons made using only the recommended values of the ^{19}F(p,α)^{16}O reaction rate. By considering the uncertainties of the 3 estimates of the reaction rate we find that the nucleosynthesis predictions agree the one with the others within the error bands. In any case it is evident (in particular from the lower panel on the right of figure 3) that the major uncertainties are those coming from the stellar observations.

5. References

[1] Lugaro M., Ugalde C., Karakas A. I., et al. 2004 *ApJ* **615** 934
[2] Cristallo S., Di Leva A., Imbriani G., et al. 2014 *A&A* **570** 46
[3] Woosley S. E., & Haxton W. C. 1988 *Nature* **334** 45
[4] Meynet G. & Arnould M. 2000 *A&A* **355** 176
[5] Palacios A., Arnould M., & Meynet, G. 2005 *A&A* **443** 243
[6] Longland R., Lorén-Aguilar P., José J., et al. 2011*ApJL* **737** 34
[7] Abia C., Cunha K., Cristallo S., et al. 2015 *A&A* **581** 88
[8] Jönsson H., Ryde N., Spitoni E., et al. 2017 *ApJ* **835** 50
[9] Cristallo S., Straniero O., Gallino R., et al. 2009 *ApJ* **696** 797
[10] Forestini M., Goriely S., Jorissen A., & Arnould, M. 1992 *A&A* **261** 157
[11] La Cognata M., Palmerini S., Spitaleri C., et al. 2015 *ApJ* **805** 128
[12] La Cognata, M., Mukhamedzhanov, A. M., Spitaleri, C., et al. 2011 *ApJL* **739** 54
[13] Angulo C., Arnould M., Rayet M., et al. 1999 *NuPhA* **656** 3
[14] Spitaleri C., La Cognata M., Lamia L., et al. 2016 *EPJA* **52** 77
[15] Indelicato I., La Cognata M., Spitaler C., et al. 2017 *ApJ* **845** 19
[16] Lombardo I., Dell'Aquila D., Di Leva A., et al. 2015 *PhLB***748** 178
[17] Ugalde C., Azuma R. E., Couture A., et al. 2008 *PhRvC* **77** 035801
[18] Pizzone R. G., D'Agata G., La Cognata M., et al. 2017 *ApJ* **836** 57
[19] D'Agata G., Pizzone R.G., La Cognata M, et al. 1992 *ApJ* **860** 61
[20] Trippella O., Busso M., Maiorca E., et al. 2014 *ApJ* **787** 41
[21] Palmerini S., La Cognata M., Cristallo S., & Busso M. 2011 *ApJ* **729** 3
[22] Sackmann I. J., & Boothroyd A. I. 1992 *ApJ* **392** L71
[23] Trippella O., Busso M., Palmerini s., et al. 2016 *ApJ* **818** 125
[24] Palmerini S., Trippella O., Busso M., et al. 2018 *Ge.Co.A.* **221** 21
[25] Palmerini S., Cristallo S., Busso M., et al. 2011 *ApJ* **741** 26
[26] Palmerini S., Trippella O., Busso M. 2017 *MNRAS* **467** 1193

[1] The extra-mixing model we use is the one described in [26]

IOP Conf. Series: Journal of Physics: Conf. Series **1308** (2019) 012017 doi:10.1088/1742-6596/1308/1/012017

Simulation study of the $\bar{\mathrm{p}}\mathrm{p} \to \bar{\Sigma}^0\Lambda$ reaction with PANDA at FAIR

G. Perez-Andrade, Karin Schönning, W.I. Andersson, Jenny Regina and Michael Papenbrock on behalf of the PANDA Collaboration

Dept. of Physics and Astronomy, Uppsala University, Uppsala, Sweden

E-mail: gabipa_28@hotmail.com

Abstract. The PANDA experiment will be one of the pillars of the new Facility for Antiproton and Ion Research (FAIR), which is currently under construction in Darmstadt, Germany. PANDA will be a fixed-target experiment which will allow the study of non-perturbative phenomena of the strong interaction. These will be probed in antiproton-proton collisions in the momentum range of 1.5 - 15 GeV/c. Within the PANDA physics program, strangeness production will be addressed through $\bar{p}p \to \bar{Y}Y$ processes. Measurements of the $\bar{p}p \to \bar{\Sigma}^0\Lambda$ channel for its comparison with the existing data of the $\bar{\Lambda}\Lambda$ channel are highly encouraged to study the role of isospin symmetry in hadron production dynamics. This work consists on a simulation study focusing on the feasibility of measuring the $\bar{p}p \to \bar{\Sigma}^0\Lambda$ channel at PANDA.

1. Introduction

Within the Standard Model (SM), a coherent understanding of the strong interaction at low energies is still missing. This represents a challenge given the striking nature of the strong coupling α_s: while being weaker at small distances, it becomes stronger at large distances. In Figure 1, a summary of α_s predictions together with experimental measurements is displayed. A good agreement between theory and experiments is seen at large momentum transfers [1], and in this region perturbative Quantum Chromodynamics (pQCD) succeeds in describing α_s. However, as one goes down in energy the strong coupling grows and pQCD predicts an unphysical divergent value for α_s [2]. The fact that different strategies have to be used to describe strong interactions at different energies calls for new investigations. One of the challenges are the different degrees of freedom that have to be taken into account at the different energy ranges: quarks and gluons are the relevant the degrees of freedom at high energies, whereas the hadrons are at low energies. At the intermediate energy scale it remains unclear which are the proper degrees of freedom to take into account. The lack of understanding of the strong interaction reflects in the poorly understood nucleons: their mass [3], spin [4], radius [5] and structure [6] are still unsolved puzzles. For instance, only about 2% of their mass is generated by the Higgs mechanism, while the rest is generated dynamically by the strong interaction. However, the exact mechanism is one of the open questions in modern physics.

IOP Conf. Series: Journal of Physics: Conf. Series **1308** (2019) 012017 doi:10.1088/1742-6596/1308/1/012017

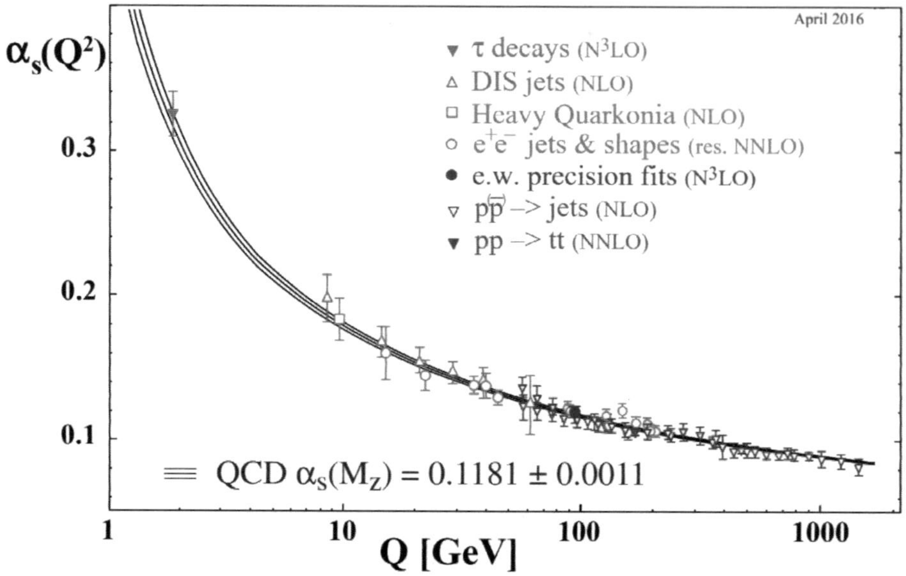

Figure 1. Summary of measurements of as a function of the respective energy scale Q. Image extracted from [1].

1.1. Hyperons

Hyperons (antihyperons), denoted as $Y(\bar{Y})$ are baryons which contain one or several $s(\bar{s})$ quarks, *e.g.* the Λ (uds) or $\bar{\Sigma}^0$ ($\bar{u}\bar{d}\bar{s}$). One of the main motivations of studying these particles is that their production is set by the mass of the s quark, m$_s$ $\sim$ 100 MeV, which is close to the QCD cut-off or hadronization scale $\Lambda_{QCD} \sim$ 200 MeV/c^2. Therefore, to study reactions containing strange hyperons at different energies, provide tools to address the question of how are quarks bounded into hadrons in the region where perturbative QCD is not valid anymore. Hyperons have larger masses than the nucleons and are unstable. Except for the Σ^0 which decays electromagnetically, all other ground-state hyperons decay weakly. Weak decays are self-analyzing *i.e.* there is a parity violating part in the decay amplitude which causes the daughter particles to be emitted according to the initial hyperon spin vector. By measuring such asymmetry, the hyperon polarization can be reconstructed [7]. In processes such as $\bar{p}p \to \bar{Y}Y$ when the decay products of the hyperon are detected, one can in addition access to some spin correlations [8]. Spin observables such as the polarization are model dependent and therefore, constitute an important testing tool for theoretical models.

2. The $\bar{\mathbf{p}}\mathbf{p} \to \bar{\Sigma}^0\Lambda$ channel

Several reactions containing hyperons have been studied at previous experiments (Figure 2). From such measurements, the total and differential cross sections, polarization of outgoing hyperons and spin correlations have been obtained at several beam momenta. However the focus has remained on single-strange hyperon production *e.g.* the $\bar{p}p \to \bar{\Lambda}\Lambda$, at low energies.

The attempts of explaining strangeness production in $\bar{p}p \to \bar{Y}Y$ reactions have motivated the development of theoretical descriptions: i) a meson exchange model (MEX) [9], ii) a quark-gluon model (QG) [10] and iii) the combination of the two afore-mentioned approaches [11]. These models describe to some extent what has been observed in the experiments. However, fail to describe the complete spin dynamics, and there are few predictions and data about other hyperons different from Λ.

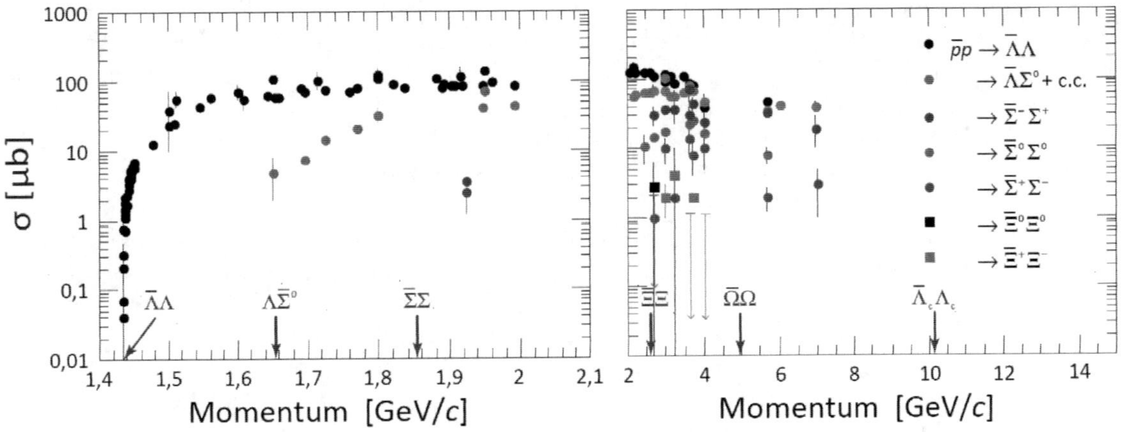

Figure 2. Hyperon production in $\bar{p}p$ annihilations [8]. Thresholds of various reactions are marked in blue.

Providing information about hyperons such as the $\bar{\Sigma}^0$, could give important clues to the underlying physics of the strong interaction mechanism. For this reason, measurements of the $\bar{\Sigma}^0\Lambda$ reaction for its comparison with the existing $\bar{\Lambda}\Lambda$ data has been motivated. The similarities in quark content between Λ and Σ^0, lead to think that these reactions dynamics can be similar. On the other hand, the isospin and spin structure should be different in the two cases. For instance, predictions given by the MEX model establish that the strength of the coupling in the $\bar{\Sigma}^0$ case is stronger than the Λ case. While, the comparison between the channels in the QG picture provides information about the isospin dependence of the production mechanism.

From previous measurements reported by the PS185 collaboration at CERN, it was concluded that the angular distributions of both the $\bar{\Lambda}$ and the $\bar{\Sigma}^0$ in all measurements of the $\bar{p}p \rightarrow \bar{\Lambda}\Lambda$ and $\bar{p}p \rightarrow \bar{\Sigma}^0\Lambda + c.c.$ reactions respectively, show a forwardly peaked behavior and that the strength of the peak depends on the beam momentum [12],[13],[14]. The existing measurements of such reactions are limited to p_{beam} up to $\sim$ 6 GeV. At higher energies, the data bank is scarce. Therefore, if one wants to get more information about strangeness production and explore the multi-strange sector, more data are needed.

3. The PANDA experiment

The antiProton ANnihilations at DArmstadt (PANDA) experiment, is one of the pillars within the Facility for Antiproton and Ions Research (FAIR) currently under construction in Germany. The PANDA experiment has a broad physics program focused on strong interaction studies. At the first stage, strangeness production will be addressed through $\bar{p}p \rightarrow \bar{Y}Y$ processes [15]. With a *Target spectrometer* and a *Forward* Spectrometer, the PANDA detector will offer a nearly 4π-solid angle coverage. Each spectrometer will contain devices to perform tasks such as particle tracking and identification, vertex detection and calorimetry. The PANDA detector will be hosted within the High Energy Storage Ring (HESR) antiproton ring. At its first stage, the HESR will provide a luminosity of 10^{31} cm^{-2}s^{-1} until the design luminosity of 1.5×10^{32} cm^{-2}s^{-1} is achieved. As part of the PANDA experiment preparations, it is of major importance to understand such reactions and study their measurement feasibility. For instance, a recent simulation study of double-strange excited Ξ^* shows that it will be possible to start collecting large and clean exclusive samples from already the first phase of operation [16]. Therefore, a high strange hyperon production rate is expected when the full luminosity is achieved.

IOP Conf. Series: Journal of Physics: Conf. Series **1308** (2019) 012017 doi:10.1088/1742-6596/1308/1/012017

4. Software tools

To perform the simulation studies, the *PandaRoot* framework is used. PandaRoot is based on ROOT [17] and Virtual MonteCarlo [18] for the detector implementation and simulation. It is specific for PANDA and has been developed to simulate the full experiment chain: from the particles generation in the $\bar{p}p$ annihilation, to the physics analysis. Several particle generators are included within PandaRoot. The most relevant for this particular study are EvtGen and Dual Parton Model (DPM) [19]. EvtGen is used to generate the signal events, including complex sequential decay channels taking into account known decay properties, such as angular distributions or polarization [15]. The DPM is a background generator in which all possible hadronic reactions are simulated at a given beam momentum and takes into account differential cross-sections parametrized from existing experimental data [20],[21]. After the event generation, the particles get transported through the PANDA detector material using GEANT [22]. Interactions such as multiple scattering and energy loss by *e.g.* ionization, [19] as well as effects from traversing magnetic fields are accounted for. After the transport, the detector responses to the particle passage are digitized. Different algorithms are used for the track reconstruction and the particle identification. These two tasks are carried out in two steps: first locally at each detector and then globally when all the single detector information are collected [23]. From the obtained information the physics analysis can be performed. With the current analysis tools implemented in PandaRoot, it is possible to obtain reconstruction efficiencies and rates for several channels containing hyperons. However, it is important to mention that PandaRoot is still under development.

5. Analysis

A study of the reaction $\bar{p}p \to \bar{\Lambda}\Sigma^0 + c.c.$ has been already performed for PANDA and it is shown in Ref. [7]. In the latter, a Software framework based on the Babar experiment [24] and an isotropic distribution for the scattering angle of the $\bar{\Sigma}^0(\Sigma^0)$ were used. Since the PandaRoot framework is now in an advanced stage of development, the reaction under study called for un update. The present work is currently under an internal review process and therefore, a general explanation of the analysis strategy will be given without showing any result. Preliminary results are presented in [25].

In this work, two samples of 10k events of the reaction $\bar{p}p \to \bar{\Sigma}^0\Lambda$ were simulated: one at $p_{beam} = 1.771$ GeV/c and one at $p_{beam} = 6$ GeV/c. These two beam momenta were chosen for easier comparison to previous measurements. Both simulations were performed using a decay model based on the results obtained by the different experiments at CERN (Refs. [12] and [14]). The $\bar{\Sigma}^0$ baryon decays into a $\bar{\Lambda}$ almost at the interaction point ($\tau_{\bar{\Sigma}^0} \approx 7.4 \times 10^{-20}$s [1]). The $\bar{\Lambda}$ and Λ decay after traveling a few centimeters ($\tau_{\Lambda/\bar{\Lambda}} \approx 2.632 \times 10^{-10}$s [1]) and thus their decay products are the particles reaching the detectors. After the generation, five final state particles per event are expected:

$$\bar{p}p \to \bar{\Sigma}^0\Lambda \to \bar{\Lambda}\gamma p\pi^- \to \bar{p}\pi^+\gamma p\pi^- \quad (1)$$

However, after the generated particles are propagated through the detector material, a non-negligible background originates and must be considered. The latter can contain the same or different particle species than the final state particles. The focus of this analysis is to test how well the $\bar{\Sigma}^0$ and Λ corresponding to the generated channel can be reconstructed and identified. This is achieved by selecting and combining their final state particles measured in the detectors. The analysis strategy is divided in a *pre-selection* and a *final selection* stage. At both stages, *kinematic fitting* procedures are used: these are statistical methods which improve the measurements of processes such as particle interaction or decay, by utilizing known information *e.g.* particle masses, energy and momentum conservation [26].

The pre-selection is divided in three steps: i) the final state particles identification, ii) a photon selection based on the reaction kinematics and iii) the $\bar{\Lambda}/\Lambda$ reconstruction through the combination of the charged candidates $p\pi^-/\bar{p}\pi^+$ respectively.

Since the main goal of the analysis is to obtain a sample where false particle candidates are as few as possible, two types of event selection methods were tested: *inclusive* or single-tag and *exclusive* or double-tag.

5.1. *Inclusive selection*

In this case, not all final state particles are detected and thus only one part of the system can be reconstructed. In this specific case, only the $\bar{p}\pi^+\gamma$ particles are detected and hence only the $\bar{\Sigma}^0$ can be directly reconstructed. The reaction is of the form $\bar{p}p \to \bar{\Sigma}^0 X$ where X denotes the undetected particle. The identity of X can be obtained from missing kinematics. Since the capability to detect a limited number of particles per event is higher, usually the inclusive selection features a good efficiency. On the other hand, fewer kinematic constraints can be imposed, meaning that the background cannot be suppressed to the same extent as if the complete system was reconstructed.

5.2. *Exclusive selection*

In this case all five final state particles are considered, so the complete $\bar{\Sigma}^0\Lambda$ system can be reconstructed. As a consequence, more information is available and the kinematic properties given by the initial state, *i.e.* the $\bar{p}p$ system, constrain the final state. The latter can be used to improve the momentum or energy resolution of the reconstructed particles. The disadvantage in this type of event selection is that the probability of particles escaping the detector is higher and thus the reconstruction efficiency is usually lower than the case of inclusive selection. In addition, since the approach implies selection of a specific decay channel, the selected sample size is reduced with the corresponding branching fraction [25].

With this analysis strategy, signal and background reconstruction efficiencies for the $\bar{\Sigma}^0\Lambda$ channel were obtain at the two p_{beam} chosen, as well as their reconstruction rates shown at [25].

6. Summary

A complete model to explain the strong interaction at all relevant energy ranges is needed. Currently, there are many open questions about the principal characteristics of the nucleon. Reactions involving hyperons are a powerful tool probe the strong interaction at the intermediate energy region and therefore at the PANDA experiment we will seek to increase the database on hyperon reactions: exploring a wider range of p_{beam} and the multi-strange region. So far, various simulation studies have shown that the PANDA experiment is very promising for the production of several hyperon channels from already the starting phase. An updated simulation study on the reconstruction feasibility of the $\bar{p}p \to \bar{\Sigma}^0\Lambda$ channel with a realistic angular distribution has been performed.

References

[1] M. Tanabashi *et al.* [Particle Data Group] 2018, Review of Particle Physics, *Phys. Rev.* D **98** no.3, 030001.
[2] Alexandre Deur *et al.* 2016, The QCD running coupling, *Progress in Particle and Nuclear Physics* **90** 1-74.
[3] S. Scherer and M. R. Schindler 2012, Quantum chromodynamics and chiral symmetry, *Lect. Notes Phys.* **830** 1.
[4] C. A. Aidala, S. D. Bass, D. Hasch and G. K. Mallot 2013, The Spin Structure of the Nucleon, *Rev. Mod. Phys.* **85** 655
[5] R. Pohl, *et al.*, 2010, The size of the proton, *Nature*, **466**, 213.
[6] G. A. Miller 2007, Charge Density of the Neutron, *Phys. Rev. Lett.* **99** 112001
[7] S. Grape 2009, Studies of PWO Crystals and Simulations of the $\bar{p}p \to \bar{\Lambda}\Lambda, \bar{\Lambda}\Sigma^0$ reactions for the PANDA Experiment. PhD thesis, Uppsala University DiVA, id: diva2:235858.

[8] T. Johansson 2003, *Proc. Int. AIP Conf. Proc. 8th Int. Conf. on Low Energy Antiproton Physics (Italy)* p 95

[9] F. Tabakin and R. A. Eisenstein 1985, Meson Exchange Calculation of the $\bar{p}p \to \bar{\Lambda}\Lambda$ reaction *Phys. Rev.* C **31** 1857.

[10] H. R. Rubinstein and H. Snellman 1985, Dynamics of QCD in the Nonperturbative Low-energy Region, *Phys. Lett.* B**165** 187.

[11] P. G. Ortega, D. R. Entem and F. Fernandez 2011, p anti-p —¿ Lambda anti-Lambda depolarization and spin transfer in a constituent quark model, *Phys. Lett.* B **696** 352.

[12] P. D. Barnes *et al.* 1996, Measurement of the anti-p p − − − > anti-lambda Lambda and anti-p p − − − > Antisigma0 Lambda + c.c. reactions at 1.726-GeV/c and 1.771-GeV/c, *Phys. Rev.* C **54** 2831.

[13] P. D. Barnes *et al.* 1990, Measurement of the Reaction $\bar{p}p \to \bar{\ell}$ambda Sigma0 + *c.c.* At 1.695-GeV/*c*, *Phys. Lett.* B **246** 273.

[14] H. Becker *et al.* [CERN-Munich Collaboration] 1978, Measurement of the Reactions $\bar{P}P \to \bar{\Lambda}\Lambda$, $\bar{p}p \to \bar{\Lambda}\Sigma^0$ and $\bar{P}P \to \bar{\Lambda}$ (Missing Mass) at 6-GeV, *Nucl. Phys.* B **141** 48.

[15] M. F. M. Lutz *et al.* [PANDA Collaboration] 2009, Physics Performance Report for PANDA: Strong Interaction Studies with Antiprotons

[16] Jennifer Pütz *et al* 2016, Study of Excited Ξ Baryons in pp-Collisions with the PANDA Detector *J. Phys.: Conf. Ser.* **742** 012028

[17] Brun, R. *et al* (1997), ROOT-an object oriented data analysis framework. *Nuclear Instruments and Methods in Physics Research Section A: Accelerators, Spectrometers, Detectors and Associated Equipment* **389**(1-2), pp 81-86

[18] Virtual Monte Carlo, `https://root.cern.ch/root/vmc/`

[19] S. Spataro [PANDA Collaboration] 2011, The PandaRoot framework for simulation, reconstruction and analysis, *J. Phys. Conf. Ser.* **331** 032031.

[20] K. Gtzen 2014, $\bar{p}p$ Cross-section estimator `https://root.cern.ch/root/vmc/`

[21] V. Flaminio *et al.* 1984, Compilation of cross-sections: p and p bar induced reactions CERN/HERA 84-01

[22] M. Asai 2006, Geant4-a simulation toolkit, *Trans. Amer. Nucl. Soc.* **95** 757.

[23] S. Spataro [PANDA Collaboration] 2008, Simulation and event reconstruction inside the PandaRoot framework, *J. Phys. Conf. Ser.* **119** 032035.

[24] G. Hamel de Monchenault 2003, The BaBar Computing Model. Stanford Linear Accelerator Center, Menlo Park, CA (US), No. SLAC-PUB-9964.

[25] G. Pérez Andrade, *Production of the* $\bar{\Sigma}^0$ *hyperon in the PANDA experiment at FAIR*, Master thesis, Uppsala University (2019) DiVA, id: diva2:1284134.

[26] P. Avery 1999, Applied Fitting Theory VI: Formulas for Kinematic Fitting. CBX 98-37

[27] S. Spataro 2012, Event Reconstruction in the PandaRoot framework. *Journal of Physics: Conference Series.* Vol. **396.** No. 2. IOP Publishing.

IOP Conf. Series: Journal of Physics: Conf. Series **1308** (2019) 012018 doi:10.1088/1742-6596/1308/1/012018

Beta decay of the Tz=-2 nucleus 64Se and its descendants

B. Rubio[1], P. Aguilera[1,2], F. Molina[2], J. Agramunt[1], A. Algora[1], V. Guadilla[1], A. Montaner-Pizá[1], A.I. Morales[1], S.E.A Orrigo[1], W. Gelletly[1,3], B. Blank[4], P. Asher[4], M. Gerbaux[4], J. Giovinazzo[4], S. Grevy[4], T. Kurtukian[4], C. Magron[4], J. Chiba[5], D. Nishimura[5], H. Oikawa[5], Y. Takei[5], S. Yang[5], D. S. Ahn[6], P. Doornenbal[6], N. Fukuda[6], N. Inabe[6], G. Kiss[6], T. Kubo[6], S. Kubono[6], S. Nishimura[6], Y. Shimizu[6], C. Sidong[6], P.-A. Söderström[6], T. Sumikama[6], H. Suzuki[6], H. Takeda[6], P. Vi[6], J. Wu[6], Y. Fujita[7], M. Tanaka[7], F. Diel[8], D. Lubos[9], G. de Angelis[10], D. Napoli[10], C. Borcea[11], A. Boso[12], R. B. Cakirli[13], E. Ganioglu[13], G de France[14], S. Go[15]

1 IFIC (CSIC- Univ. Valencia), 2 CCHEN-Chilean Nuclear Energy Commision, 3 Surrey, University, 4 Centre d'Etudes Nucl\'eaires de Bordeaux-Gradignan, 5 Department of Physics, Tokyo City University, 6 RIKEN Nishina Center, 7 Dep. of Physics Osaka University, 8 Institute of Physics Universität zu Köln, 9 Physics Department TU München, 10 LNL INFN Legnaro, 11 National Institute for Physics and Nuclear Engineering IFIN HH, 12 Instituto Nazionale de Fisica Nuclear, Univ. Padova, 13 Department of Physics, Istanbul University, 14 GANIL, 15 Department of Physics Tenessee University

Berta.rubio@ific.uv.es

Abstract. The beta decay of the Tz=-2 nucleus 64Se has been studied in a fragmentation reaction at RIKEN-Nishina Center. 64Se is the heavies Tz=-2 nucleus that decays to bound states in the daughter nucleus and the heaviest case where the mirror reaction 64Zn(3He,t)64Ga on the Tz=+2 64Zn stable target exists and can be compared. Beta-delayed gamma and proton radiation is reported for the 64Se and 64As cases. New levels have been observed in 64As, 64Ge (N=Z), 63Ge and 63Ga. The associated T1/2 values have been obtained.

1. Introduction

This paper concerns the study of proton rich nuclei which together with their mirror partners can provide information on isospin symmetry far from the stability. If isospin symmetry works mirror nuclei should have identical structure, just exchanging the number of protons by the number of neutrons, and mirror processes such as charge exchange reactions and beta decay should produce similar results. Based on these ideas we have launched a programme of beta decay experiments at GSI [1], GANIL [2,3] and RIKEN where we have studied the decay of Tz=-1 and -2 nuclei in the *fp* shell produced in fragmentation and compared our results with those obtained in (3He,t) charge exchange reaction experiments on the mirror stable nuclei carried out at RCNP Osaka (see for instance [4]). The first experiments were carried out at GSI and GANIL using either 58Ni or 64Zn beams. In these

IOP Conf. Series: Journal of Physics: Conf. Series **1308** (2019) 012018 doi:10.1088/1742-6596/1308/1/012018

studies, the heaviest nuclei which could be reached were 56Zn [2] and 58Zn [5]. In order to reach 60,62Ge and 64,66Se, an intense beam of 78Kr, only available at the RIKEN Nishina center, was necessary. Here we present preliminary results from the RIKEN campaign on the decay of 64Se and its descendants.

2. The experiment

The nuclei of interest were produced at the RIKEN Nishina Center using the fragmentation of a 345 MeV 78Kr beam with unprecedented intensity (up to 300 particle nA) on a Be target. The fragments were separated in flight using the BigRIPS separator [6] and implanted in three 1mm thick WAS3ABi double-sided Si strip detectors (DSSSD) [7] each of them having an active area of 60 x 40 mm2 segmented into 60 vertical by 40 horizontal strips. The implantation setup was surrounded by the EUROBALL-RIKEN Cluster Array (EURICA) [8], 12 CLUSTER-detectors of Euroball type, each one consisting of seven tapered hexagonal HPGe crystals at a nominal distance of 22 cm from the center of WAS3ABi. Three different settings of BigRIPS were used in this experiment, the first two were optimised for the production of 60Ge, 62Ge and 64Se nuclei and the third one for 66Se. Putting conditions on the ToF and DeltaE signals of the ions we have produced the identification plots corresponding to these three settings and they are shown in Figure 1 and table 1.

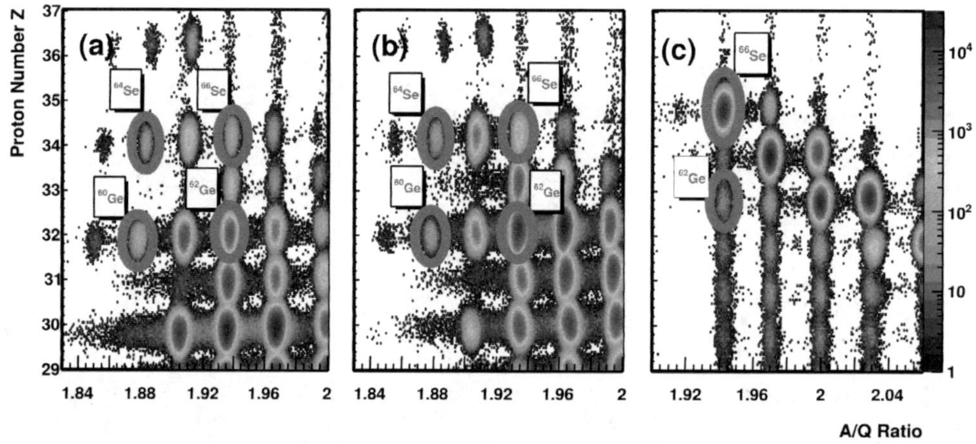

Figure 1. Particle identification plot for the three Big-RIPS settings used at Nishina Center–RIKEN. The first two, (a) and (b) were optimised for the production of 64Se, 62Ge and 60Ge, and (c) was set for 66Se.

Nucl.	Number of Implanted Ions per setting			Total Implanted
^{64}Se	3829	6479	-	10308
^{64}As	23446	457474	976	481896
^{63}Ge	117059	3741115	2798	3860972
^{66}Se	11538	78634	952026	1042198

Table 1: Number of implanted ions for the nuclei of interest in the three settings used in the experiments.

3. Results for the decay of 64Se and its descendants

IOP Conf. Series: Journal of Physics: Conf. Series **1308** (2019) 012018 doi:10.1088/1742-6596/1308/1/012018

The 64Se nucleus is an interesting case to study. First of all, it is the heaviest Tz=-2 case where the beta decay can be compared with the charge exchange mirror reaction (3He,t) on the 64Zn stable nucleus [9], secondly it is the heaviest Tz=-2 nucleus where the decay still proceeds through beta-delayed protons and beta-delayed gamma radiation, thirdly, it is only one neutron away from the last bound selenium isotope, 63Se, a two-proton emitter [10]. Moreover, the beta-delayed proton decay can populate excited states in the Tz=-1/2 nucleus 63Ge, a nucleus that cannot be accessed directly by beta decay because 63As is unbound.

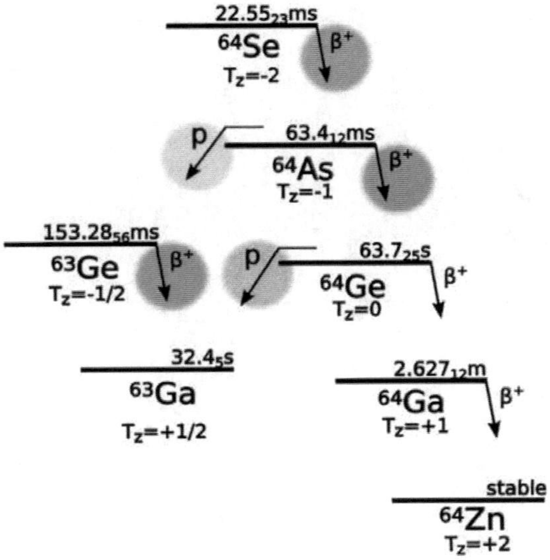

Figure 2. The decay chain associated with 64Se decay.

No experimental information on the decay properties of 64Se existed prior to this work. In consequence, a full understanding of all the decays correlated with the implantation of 64Se ions was mandatory in order to deduce the primary radiation due to 64Se decay. The decay chain associated with 64Se is shown in figure 2 where decays shorter than 200 ms are marked in color. A strong beta-delayed proton branching is expected in 64Se since the proton separation energy in the daughter nucleus 64As is expected to be of the order of 100 keV according to the last Mass Evaluation tables [11]. In consequence, three beta decays are important here, the nucleus of interest, 64Se, and the descendants, 64As and 63Ge. The gamma spectra associated with the three corresponding implants are presented in figure 3. The gamma radiation is detected in EURICA, the spectra are constructed in add-back mode in prompt coincidence with the beta-particles which happen in time correlation with the implantation conditions. In green, we show the gamma lines that are observed in time correlation with 64Se implants but not with 64As or 63Ge, they are associated with the beta decay of 64Se. Similarly, the lines marked in pink appear with the condition on 64Se (as daughter activity) and 64As, but disappear when the condition is set on 63Ge, and are thus associated with the beta decay of 64As. Finally, the gamma lines associated with the decay of 63Ge are marked in red. No beta-delayed gamma radiation was known in any of these three decays prior to this work.

IOP Conf. Series: Journal of Physics: Conf. Series **1308** (2019) 012018 doi:10.1088/1742-6596/1308/1/012018

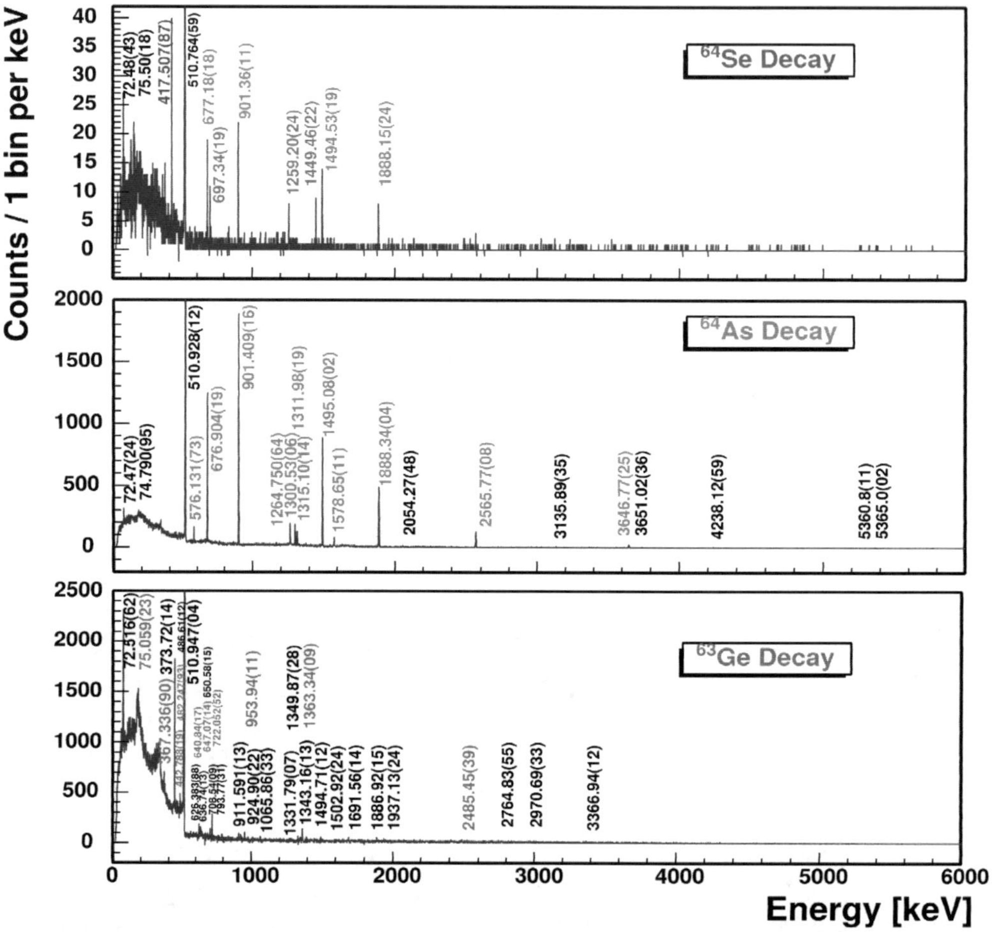

Figure 3. From top to bottom, the gamma spectra associated with the implantation of 64Se, 64As and 63Ge. The correlation time windows are 170, 440 and 1050 ms respectively. They have been used to deduce the origin of the lines observed in correlation with 64Se implantations. Gamma energies marked in green, pink and read are assigned to the decay of 64Se, 64As and 63Ge respectively.

Beta-delayed proton decay was expected to happen with high probability in the decay of 64Se because of the low proton separation energy and with lower probability in the decay of 64As. These expectations were confirmed as can be seen in the corresponding proton spectra shown in figure 4. They are constructed using time correlations between the implanted ions and the decay signals (betas and protons) detected in the same pixel defined as the crossing between one X and one Y strip. The signal below 1 MeV corresponds to the deltaE beta signal. Peaks above 1 MeV originate from the proton decay, in general summed with the beta signal. The proton peak centroids were obtained by fitting the spectra with the simulated beta-delayed proton response using the code GEANT4. The WAS3ABi DSSSD detectors were calibrated using an electron conversion source of 207Bi; thus energy shifts are expected due to the several absorption effects before the electron penetrates into the detector and to the summing of the protons with the betas. To take into account all these effects an internal calibration is essential. The energy shifts were evaluated using 57Zn and 71Ge beta-delayed proton spectra obtained in the same experiment and constructed in the same manner as the 64Se beta-

IOP Conf. Series: Journal of Physics: Conf. Series **1308** (2019) 012018 doi:10.1088/1742-6596/1308/1/012018

delayed proton spectrum. The corresponding energies were taken from [10]. The uncertainties in the proton energies shown in figure 4 originate from the fit of the response function as well as the systematic shifts of the proton energies compared with those reported in the literature [10].

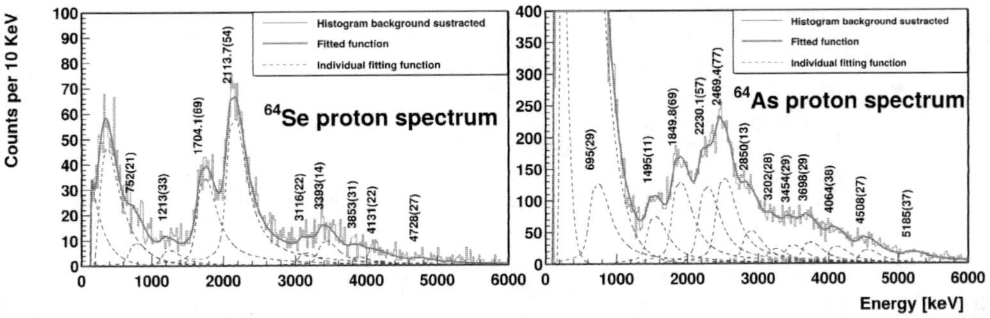

Figure 4. DSSSD spectra associated with the beta decays of 64Se and 64As. Peaks associated with beta-delayed proton decay are marked with energies.

The half-lives of the three nuclei under study are presented in figure 5. They are based on the correlations between the implanted 64Se and 64As ions and the proton decays happening before and after the implantation in the same pixel, and the implantation-beta correlations in the case of 63Ge.

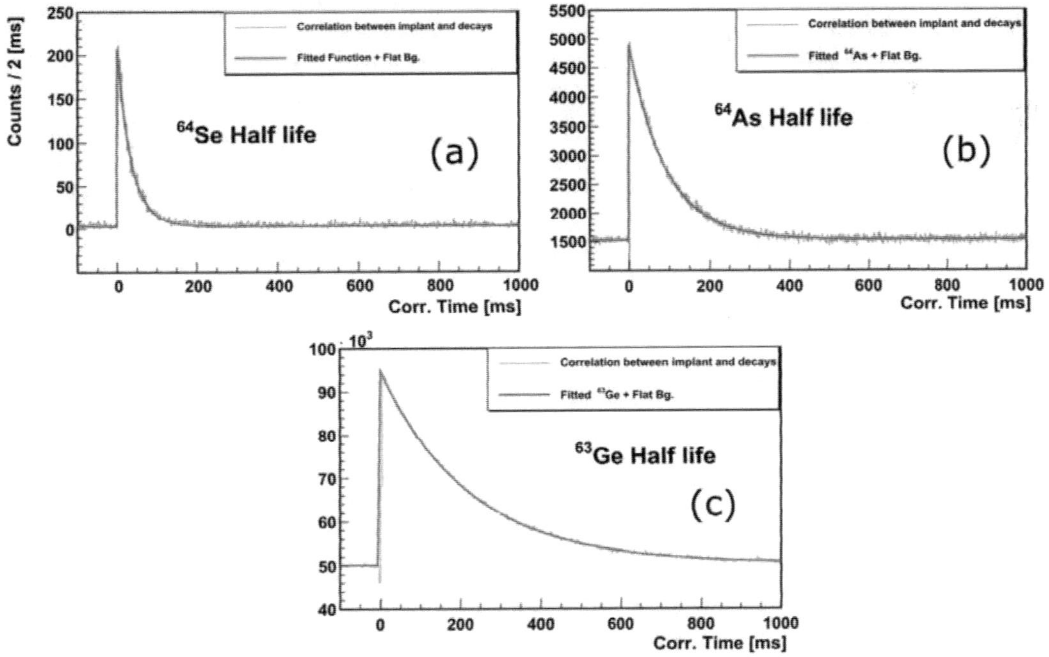

Figure. 5: Fit of the implant-beta and implant-proton correlation for the half-life of (a) 64Se, (b) 64As, (c) and 63Ge.

IOP Conf. Series: Journal of Physics: Conf. Series **1308** (2019) 012018 doi:10.1088/1742-6596/1308/1/012018

Based on the analysis of all these data, including gamma–gamma as well as proton-gamma coincidences we have obtained the following preliminary level schemes:

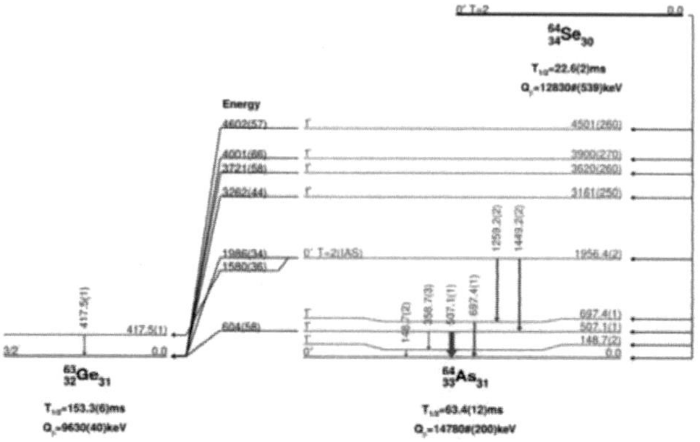

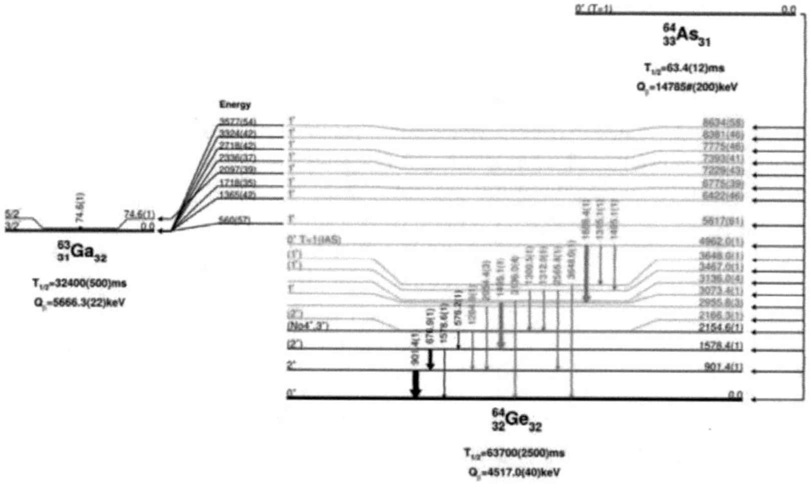

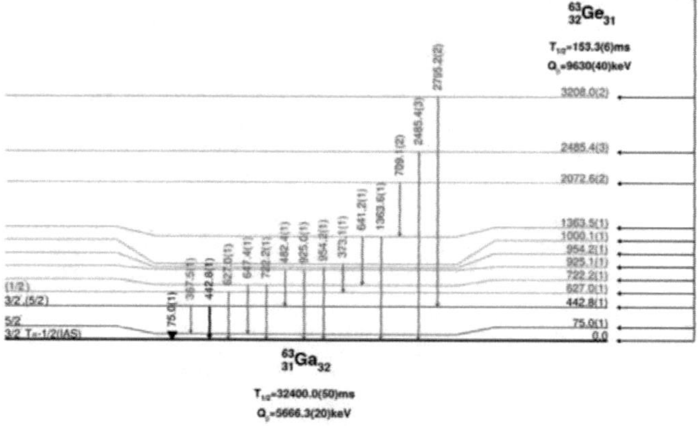

IOP Conf. Series: Journal of Physics: Conf. Series **1308** (2019) 012018 doi:10.1088/1742-6596/1308/1/012018

Figure 6. Decay schemes for 64Se, 64As and 63Ge studied in this work. Colors represent the gamma lines observed for the first time in this work, while the ones previously reported in the literature are shown in black. In the case of 64Ge, they were reported in the in-beam work of Farnea et al. [12], and 63Ge data are from [13,14]. The Q-beta values are taken from [11].

4. Discussion

As mentioned in the introduction the study of the decay of 64Se was one of the main goals of the present experiment. A strong population of the IAS at around 2 MeV is expected in the decay. This level is identified as the experimental level at 1955 keV energy. The IAS has isospin 2 and spin parity 0+. Although this level is well above the proton separation energy, it decays by proton emission as well as by electromagnetic transitions. This is due to the forbidden character of this proton decay since it originates from a state with T=2 and populates states in 63Ge with (presumably) isospin ½. Two gamma cascades de-excite the IAS and define its excitation energy. The direct transition from the state at 506 keV to the ground state is difficult to observe because it is obscured by the 511 keV annihilation peak. Four other states are strongly populated in the decay of 64Se and are consequently assigned as 1+ levels. The lack of evidence of direct beta feeding from 64Se to the ground state of 64As, together with the information from the mirror nucleus of the 0+ character of the 64Zn ground state, defines the ground state of 64As as 0+. Such a state is normally called the AAS (Anti Analogue State) and it is expected to have a configuration very similar to the IAS but with a different isospin.

The decay scheme of 64Se can be directly compared with the charge exchange reaction on the mirror stable nucleus 64Zn [9], the energies of the relevant levels can be compared in table 2. Good isospin symmetry is observed. A more detailed comparison of the Gamow-Teller strength B_{GT} and Fermi strength B_F is in preparation [15].

Similarly, a good isospin symmetry is observed for the T=1/2 mirror pairs 63Ge and 63Ga.

Ex. 64As (keV)	Ex. 64Ga (keV)	Ip
0	0	0+
147	127	1+
506	426	1+
697	666	1+
1955	1923	0+
Ex 63Ge (keV)	Ex 63Ga	Ip
0	0	3/2-
75	75	5/2-
417	443	

Table 2: Comparison of mirror states in 64As and 64Ga, and in 63Ge and 63Ga. The excitation energies originate from the present study for 64As, 63Ge and 63Ga and from [9] for 64Ga. As can be seen, good isospin symmetry is observed.

Finally, new excited states could be observed in the N=Z nucleus 64Ge populated in the decay of 64As. Nuclei with N=Z are particularly interesting because proton-neutron correlations are supposed to play an important role in their structure. Moreover N=Z even-even nuclei can provide information

IOP Conf. Series: Journal of Physics: Conf. Series **1308** (2019) 012018 doi:10.1088/1742-6596/1308/1/012018

on isospin symmetry. For instance, in 64Ge, a study by Farnea et al [12] using a fusion evaporation reaction revealed a forbidden E1 transition which could only be explained in terms of T=0 and T=1 mixing. In this in-beam study a number of yrast states were observed up to spin 14+. In our study, we could locate a number of low spin states, mainly 1+ states as well as the 0+ IAS of 64As located at 4961 keV excitation energy.

References

[1] F. Molina et al. Phys. Rev. C 91, 014301 (2015)
[2] S. E. A. Orrigo et al. Phys. Rev. Lett 112, 222501 (2014)
[3] S. E. A. Orrigo et al. Phys. Rev. C 93, 044336 (2016)
[4] Y. Fujita, B. Rubio, and W. Gelletly, Prog. Part. Nucl. Phys. 66, 549 (2011)
[5] L. Kucuk et al. Eur. Phys. Jour. A 53, 134 (2017)
[6] T. Kubo et al., Prog. Theor. Exp. Phys. 2012, 03C003 (2012)
[7] S. Nishimura et al., RIKEN Accel. Progr. Rep. 46, 182 (2013)
[8] P.-A. Söderström et al., Nucl. Instr. and Meth. in Physics Research B 317, 649652 (2013)
[9] F. Diel et al Phys. Rev. C 99, 054322 (2019)
[10] Blank et al. Eur. Phys. Jour. A 31, 267 (2007)
[11] AME2016, http://amdc.in2p3.fr/web/masseval.html
[12] E. Farnea et al. Phys. Lett B 551, 56 (2003)
[13] D. P. Balamuth et al. PRC 43, 5 (1991) 2082-2097
[14] M. Weiszflog et al. Eur. Phys. Jour. A 11, 25 (2001)
[15] P. Aguilera Ph.D Thesis Universidad de Santiago de Chile. 2019

IOP Conf. Series: Journal of Physics: Conf. Series **1308** (2019) 012019 doi:10.1088/1742-6596/1308/1/012019

Observation of fission events in the ^{14}N+^{181}Ta system at projectile energies ≈ 5-6 MeV/A

Vijay R Sharma[1], R Kumar[2,⋆], B P Singh[3,†], E F Aguilera[1], Mohd Shuaib[3], S Mukherjee[4], R Dubey[5], P P Singh[6], S Kumar[2], S Appannababu[9], Abhishek Yadav[7], M K Sharma[8], J C Morales-Rivera[1], and R Prasad[2]

[1]Departamento de Aceleradores, Instituto Nacional de Investigaciones Nucleares, Apartado Postal 18-1027, Caixa Postal 11801, Distrito Federal, México
[2]NP-Group, Inter-University Accelerator Centre, New Delhi-110 067, India
[3]Department of Physics, Aligarh Muslim University, Aligarh 202 002, India
[4]Physics Department, Maharaja Sayajirao University of Baroda, Vadodara 390002, India
[5]iThemba LABS, National Research Foundation, P.O.Box 722, 7129 Somerset West, S. Africa
[6]Department of Physics, Indian Institute of Technology Ropar, Punjab-140 001, India
[7]Department of Physics, Jamia Millia Islamia, New Delhi-110025, India
[8]Department of Physics, Shri Varsheny College, Aligarh, Uttar Pradesh 202001, India
[9]Departamento de Fisica Nuclear, IF-Universidade de Sao Paulo, CEP 05508 090, Brazil

E-mail: ⋆rakuiuac@gmail.com, and †bpsinghamu@gmail.com

Abstract. Present paper reports the mass distribution of fission events in the ^{14}N+^{181}Ta reaction at four different projectile energies viz., 82.2 ± 0.8, 79.18 ± 0.82, 76.8 ± 1.2 and 72.9 ± 0.91 MeV using recoil catcher technique followed by off-line γ-ray spectrometry. A single peaked broad Gaussian mass curve has confirmed the nonappearance of any non-compound nuclear fission. Further, the variance of the mass distribution is compared with the existing literature data. It has been observed that the lower mass asymmetric system results in a lower variance of the mass distribution.

1. Introduction

The nuclear fusion process is defined as the capture of a projectile by the target nucleus forming fully equilibrated compound nucleus which subsequently decays by emission of particles followed by γ-rays to form evaporation residues (ERs). When the compound nucleus is heavy the fission process competes strongly with the evaporation of particles at each stage of the evaporation process, however, the role of the angular momentum (ℓ) transmitted to the target nucleus is also essential. For the values of ℓ above the critical angular momentum (ℓ_{crit}), the fission barrier decreases and an immediate fission process route may take place. This is also a limiting factor in the production of superheavy nuclei elements [1].

Recent experimental data on fusion [2, 3, 4, 5] at projectile energies 15% to 25% above the Coulomb barrier suggests the presence of fission events involving the projectiles such as ^{12}C,^{14}N,^{16}O and ^{20}Ne on medium mass targets, where fusion is expected to be a dominant mode of reaction process. In-spite of extensive work carried out for a wide range of excitation energy, and other entrance channel parameters [4, 6, 7, 8] a complete understanding of the

Figure 1. A typical snapshot of the General Purpose Scattering Chamber (GPSC) (a) inside view, target ladder is shown in the inset, (b) full view, and (c) In-vacuum transfer facility (ITF) setup view. For details see text.

mechanism of various types of reactions populated at low energies is still missing. Measurement of evaporation residues are signature of compound nucleus formation and useful probe to study the statistical as well as dynamical aspects of fusion-fission reactions.

For gaining a better insight into heavy ion reactions, a detailed study of the decay products of the compound nucleus ^{195}Hg in ^{14}N+^{181}Ta system has been undertaken by our group. A part of the data analysis involving fission events is reported in this paper.

2. Experimental Details

The experiment has been performed using $^{14}N^{6+,7+}$ beams from the 15UD pelletron accelerator at the IUAC, New Delhi, India. The thin Ta target foils ($\approx$ 99.9 %) of 1.3 to 1.9 mg/cm^2 and Al catchers of $\approx$ 1.4 to 2.0 mg/cm^2 were prepared using the rolling technique. Each target was followed by Al-catcher. Irradiations were performed in the GPSC of 1.5 m diameter having an in-vacuum transfer facility (ITF) using conventional recoil catcher technique. Using this ITF facility the samples after irradiation may be changed in the GPSC without disturbing the vacuum inside the chamber. Thus, the time lapse between the stop of the irradiation and the counting of the samples may be considerably reduced and thus induced activities of short half-lives may be recorded. Keeping in view the half-lives of interest ranging from few minutes to several hours, irradiations have been carried out for $\approx$ 8-10 hours. Typical photographs of the GPSC are given in figure 1 (a) inner view, target ladder in the inset, figure 1 (b) a full view of GPSC and figure 1 (c) the ITF setup. The flux of the incident ^{14}N ions was monitored using an ORTEC current integrator. The samples of ^{181}Ta alongwith appropriate catcher foil were irradiated at 82.2 $\pm$ 0.8, 79.18 $\pm$ 0.82, 76.8 $\pm$ 1.2, and 72.9 $\pm$ 0.91 MeV beam energies at a constant beam current $\approx$ 25 pnA. The activities produced in the samples were recorded off-line by HPGe detector of 100 c.c. active volume coupled to a CAMAC based software CANDLE [10]. The detector used in this experiment was pre-calibrated for energy and efficiency using various standard γ-sources viz., ^{60}Co, ^{133}Ba and ^{152}Eu at different source-detector separations. A typical plot of the photo peak efficiency of HPGe detector as a function of γ-ray energies varying from 121 keV to 1408 keV using ^{152}Eu point source at source-detector distance 3.0 cm, is shown in Fig. 2. It may be mentioned that the target-detector separation was suitably adjusted so as to keep the dead time $<$ 10 %. In order to detect and follow the residues of longer half-lives, the counting of irradiated samples has been done for several days. In the present work, we identified 22 fission events based on their characteristic energy of γ-lines and also from their measured half lives. For the sake of completeness a typical decay curve plot for an arsenic radio-isotope (^{70}As) is shown in figure 3. The measured half-lives of all the fission-like residues [13] were found to be in good agreement with their literature values [11].

IOP Conf. Series: Journal of Physics: Conf. Series **1308** (2019) 012019 doi:10.1088/1742-6596/1308/1/012019

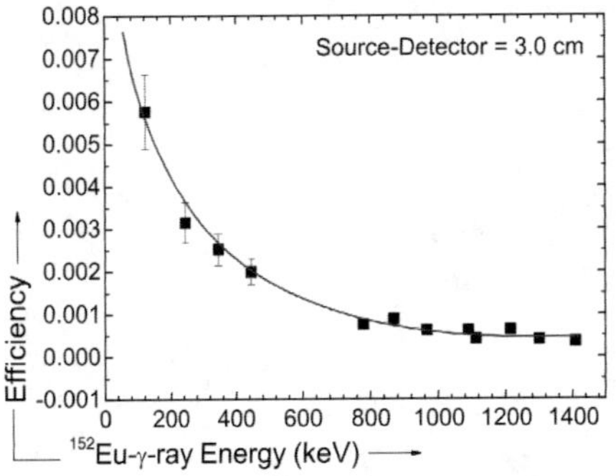

Figure 2. Typical plot of photo peak efficiency of HPGe detector as a function of γ-ray energy. The solid line is drawn to guide the eye.

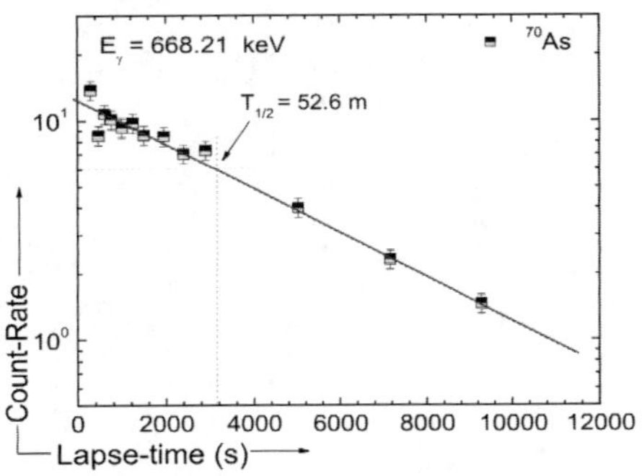

Figure 3. Typical decay curve of arsenic residue (^{70}As) at projectile energy 82.2 ± 0.8 MeV.

3. Data Reduction, Results and their interpretations

An experiment on the ^{14}N+^{181}Ta reaction suggesting the population of radio nuclide viz., ^{191}Hg, ^{190}Hg, ^{191}Au, ^{190}Au, ^{188}Pt, and ^{187}Pt via complete fusion (CF) and/or incomplete fusion (ICF) processes in the framework of theoretical model code PACE [12] has been reported recently [13]. Further analysis of the experimental data done in the present work revealed the presence of several residues which are not expected to be populated either by CF and/or ICF processes. These residues were found to have charge and atomic mass values around half of the values for the residues produced by composite systems formed as a result of fusion of projectile and the target nucleus, indicating the possibility of their production only through fission of composite systems. In a qualitative way, the fission arising due to the decay of the excited composite system formed via complete momentum transfer from projectile to the target nucleus may be termed as complete fusion-fission and via incomplete momentum transfer from the projectile to the target nucleus may be termed as incomplete fusion-fission. Though the identification of 22 fission-like events based on decay curve analysis as has been explained in the previous section, however, the formation cross-sections of such identified fission products is determined from the FORTRAN program EXP-SIGMA based on standard formulation described elsewhere [13] at 82.2 ± 0.8, 79.18 ± 0.82, 76.8 ± 1.2, and 72.9 ± 0.91 MeV beam energies. It may be pointed out that the overall error including statistical errors is estimated to be ≤ 15%, excluding the uncertainty in branching ratio, decay constant, etc., which have been taken from the Table of Radioactive Isotopes [11].

The plots of experimentally determined production cross-sections of various fission fragments at the studied energies are shown in figure 4 (a-d), respectively. The upward arrows indicate that only the metastable states have been measured and the total production cross-sections of these fission fragments are expected to increase. As can be seen in figure 4, the mass distributions of fission fragments are almost symmetric and can be fitted with one Gaussian function (as given in eq. 1) indicating the formation from the decay of compound nucleus.

$$Y(A) = \frac{Y_Z}{\sqrt{2\pi\sigma_A^2}} e^{-(A-A_p)^2/2\sigma_A^2} \tag{1}$$

where, the symbols have their usual meaning [14]. From eq.1, the parameter mass variance (σ^2) is obtained after the Gaussian fit and may be used to understand the behavior of stability of the decaying nucleus into fission via fissility. In order to explore this aspect, Itkis et al. [15] analyzed a large collection of data over a wide range of fissility of compound nucleus at medium

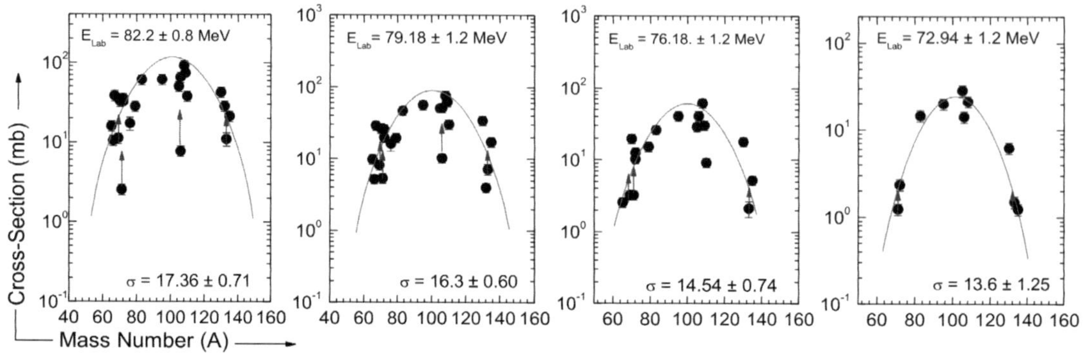

Figure 4. (Color Online) The plots of experimentally determined production cross-sections of various fission fragments at four different energies. The solid red line is the Gaussian fitting. The size of the filled circles includes the uncertainty in the yield values.

excitation energies. Moreover, the behavior of the stability (in terms of saddle point) for any system against the mass asymmetry (α) can be understood in the frame of Businaro-Gallone point (α_{BG}) [16], which is defined as the mass asymmetry for which the potential energy (saddle point) is maximum for a given fissility (χ = Z^2/A). In the present work, the mass asymmetry (α=(A_T-A_P)/(A_T+A_P)) for the presently studied system is 0.856 and is found to be greater than the critical mass asymmetry (α_{BG} = 0.827) of the system which suggests that the system will establish a mono-nuclear compact shape which facilitate equilibrium in all degrees of freedom and thus fission proceeds via compound nuclear processes and hence, the mass distribution is expected to be broad and symmetric.

Further, an attempt has been made to understand the role of entrance channel asymmetry on the behavior of mass distribution of fission-like fragments, the available mass variance of the fission fragments for different projectile-target combinations were compared with the presently calculated mass variance as a function of mass asymmetry (α). Figure 5 shows the distribution of the variance w.r.t α for four projectile-target combinations at constant projectile energy normalized with the Coulomb barrier (V_B). From the figure, it is noticed that the mass variance increases with the mass asymmetry of the interacting ions. This suggests a broader distribution of fission fragments for a more mass asymmetric system. However, more projectile-target combinations with different mass asymmetry values are needed to understand the dependence of mass variance on α.

In order to track the change in mass variance (σ^2) with excitation energy, the value of σ^2 obtained from the Gaussian fitting procedure of mass distributions of fission-like fragments is plotted as a function of excitation energy in figure 6. As can be seen in figure 6, the value of σ^2 increases with excitation energy, indicating larger spread in fission-fragment masses for higher excitation energies. The observed variation in the value of σ^2 with excitation energy for the present system follows the same trend as that reported by Ghosh et al. [19] in the above-barrier region for three different projectiles (^{19}F, ^{16}O, and ^{12}C) on a deformed target ^{232}Th (see figure 4(a-c) of Ref [19]). It may be mentioned that the present system was studied at only four energies above the barrier, hence the variation of the value of σ^2 with excitation energy at and below the barrier energies needs to be further investigated to better understand this aspect.

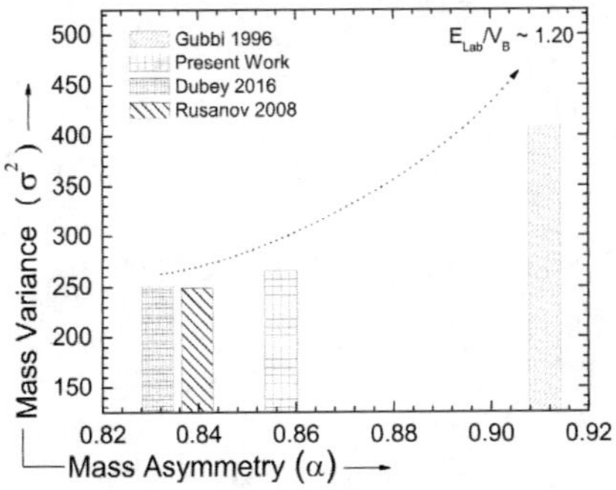

Figure 5. (Color Online) Mass asymmetry vs variances for the three projectile-target combinations (Gubbi 1996 [14], Dubey 2016 [17], Rusanov 2008 [18]). Dotted arrow is drawn to guide the eyes.

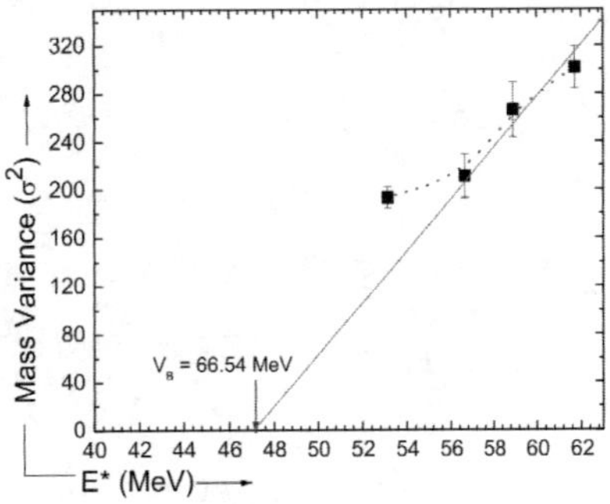

Figure 6. (Color Online) Mass variance as a function of excitation energy for ^{195}Hg. The red solid lines show the increase in σ^2 with excitation energy. Blue dotted line is drawn to guide the eyes.

4. Summary and Conclusions

The present paper reports the production cross-section of several fission fragments populated at four set of projectile energies. The mass distribution parameters obtained from the present measurements are studied in respect of mass asymmetry at constant normalized projectile energies. It has been observed that mass variance seems to exponentially increase with mass asymmetry but more experiments are required for different projectile and target combinations to explore the actual picture of the mass asymmetry systematics. The mass distribution of fission fragments at different excitation energies was studied to probe behavior of fission fragments.

5. Acknowledgments

The authors thank the Director of the IUAC, New Delhi, India for extending experimental facilities. VRS and EFA thank CONACYT for providing financial support under Project No. CB-01-254619. One of the author RD thanks SARCHI-DST, South Africa for financial support.

References

[1] Carlos A Bertulani Wiley Online Library, Nuclear reactions, 2014
[2] Tripathi R et al 2016 Phys Rev C **74** 014610
[3] Pal A et al 2018 Phys Rev C (R) **98** 031601
[4] Singh P P et al 2008 Int J Mod Phys E Vol 17, No3, 549-566
[5] Escher Jutta E et al 2012 Rev Mod Phys **84** 353-397
[6] Ngo Ch 1985 Prog Part Nucl Phys **16** 139
[7] Todd J R D et al 1993 J Phys G **19** 187
[8] Ramamurthy V S 1990 Phys Rev Lett **65** 25
[9] SRIM06: http://www.srim.org /
[10] CANDLE - Kumar B P Ajith et al 2001 DAE SNP, Kolkata
[11] Decay Radiation database version of 01 - 03 - 2018, http://www.nndc.bnl.gov/nudat2/
[12] Tarasov O B and Bazin D 2003 Nucl Instrum Methods Phys Res Sect B **204** 174
[13] Sharma V R 2019 Phys Rev C 99, 034617 (2019) and references therein
[14] Gubbi G K et al 1999 Phys Rev C **59** 3224-6
[15] Itkis M G et al 1995 Phys At Nucl **58** 2026
[16] Businaro U L and Gallone S 1995 Nuovo Cimento **1** 1277
[17] Dubey R et al 2016 Phys Lett B **752** 338-343
[18] Rusanov A Ya et al 2008 Phys At Nucl **71(6)** 956
[19] Ghosh T K et al 2005 Phys Lett B **627** 26

IOP Conf. Series: Journal of Physics: Conf. Series **1308** (2019) 012020 doi:10.1088/1742-6596/1308/1/012020

First measurements of capture reactions for the γ-process using HECTOR

A Simon[1,*], R Kelmar[1], O Olivas-Gomez[1], E Churchman[1], P Millican[1], C S Reingold[1], T Anderson[1], A M Clark[1], N Cooper[1], A C Dombos[1,2], B Frentz[1], S L Henderson[1], S Kelly[1], D Robertson[1], M Skulski[1], E Stech[1], S Y Strauss[1], W P Tan[1], B Vande Kolk[1]

[1] Department of Physics and Joint Institute for Nuclear Astrophysics, University of Notre Dame, Notre Dame, IN 46556

E-mail: *anna.simon@nd.edu

Abstract. The first measurements of the capture reactions for the astrophysical γ-process using HECTOR, a total absorption spectrometer are presented. Two reactions: ^{102}Pd(p,γ)^{103}Ag and ^{90}Zr(α,γ)^{94}Mo were measured using the γ-summing technique. The results are compared with the previous measurements found in the literature and with the NON-SMOKER predictions. The results from the current work are in a good agreement with the data found in the literature and serve as a test of the newly developed detector.

1. The γ-process nucleosynthesis

There are multiple nucleosynthesis processes responsible for production of elements heavier than iron, see Figure 1. Since these elements cannot be synthesized during stellar burning processes due to the high Coulomb barrier and thus a low reaction cross section, they have to be produced during reactions with uncharged particles [1]. There are two processes that explain the origin of most of the heavy isotopes: the s- and r-processes [2, 3, 4], which are a combination of neutron capture and β-decay. However, these processes do not explain the origin of several of the proton-rich isotopes of the elements between selenium and mercury that are shielded from the β-decay path by the valley of stability [5]. To be precise, there are 35 isotopes that cannot be produced via neutron capture processes. Several processes have been proposed to explain the origin of these nuclei [6]; however, the most favorable scenario to-date is the γ-process [7].

The γ-process is comprised of mainly γ-induced (photodisintegration) reactions: (γ,n), (γ,p) and (γ,α). The temperatures necessary for the γ-process to occur are of the order of 1.5-3.5 GK [5, 8]. As such, the γ-process has to take place in explosive environments, for example, type II [7, 8, 9, 10] or type Ia supernovae (SN) [11, 12, 13, 14, 15], where such temperatures can be achieved. In these environments, during the explosion, a high flux of γ rays is emitted. These γ rays pass through the volume of the star with the explosion shockwave and interact with the matter present in the environment (i.e., the elements produced during the stellar burning stages and by the s-process) triggering a sequence of (γ,p), (γ,α) and (γ,n) reactions. These reactions produce unstable, proton-rich nuclei that, after the flux of γ rays ends, β-decay towards stability populating the stable, proton-rich isotopes of heavy elements.

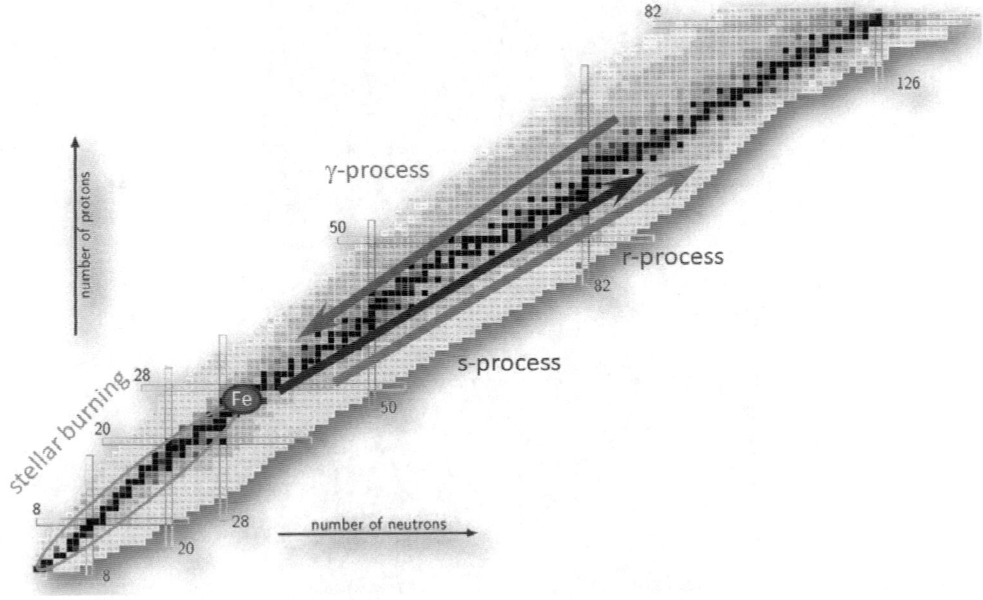

Figure 1. Nucleosynthesis processes. Black squares denote stable nuclei. Colors indicate the nuclear half-lives: blue - above 10 s, green - 1 ms to 10 s, orange - below 1 ms.

The complete network describing the γ-process involves nearly 2,000 nuclei and 20,000 reactions. Since a vast majority of those reactions cannot be measured, the models have to rely on theoretical predictions for the reaction rates, such as the statistical Hauser-Feshbach (HF) model [16]. The HF model in turn depends on the description of the statistical parameters of the nuclei, such as the level density (LD), optical model potential (OMP) and the γ-ray strength function (γSF). With a variety of models for the statistical inputs, the reaction rates obtained from HF calculations can span up to an order of magnitude, thus limiting the predictive power of the network calculations. However, experimental efforts can constrain those models by providing systematic measurements of the reaction cross sections that can then be used to constrain the inputs for the HF models.

The reactions relevant for the astrophysical γ-process can be studied via their inverse, i.e., (p,γ) and (α,γ) as the temperature range of the γ-process corresponds to projectile energies of 1-6 MeV and 3-12 MeV for proton and α capture reactions, respectively. In this work, the first results from a γ-ray summing detector at the University of Notre Dame aiming at constraining the HF predictions are presented. Cross sections for two capture reactions will be presented with a comparison to the measurements found in the literature and to the NON-SMOKER [17] calculations based on the HF model.

2. γ-summing technique

The proton and α capture cross sections depend strongly on the projectile energy and decrease rapidly by orders of magnitude within the Gamow window as the projectile energy decreases. Thus, very efficient experimental techniques are required to extend the measurements towards the low energy part of the Gamow window.

The results presented in this paper were obtained using the γ-summing technique. The technique is a well established method used for cross section measurements [18, 19, 20, 21]. It requires a large volume γ-ray detector covering almost a 4π solid angle so that all the γ rays

IOP Conf. Series: Journal of Physics: Conf. Series **1308** (2019) 012020 doi:10.1088/1742-6596/1308/1/012020

Figure 2. HECTOR: High EffiCiency TOtal absorption spectrometeR.

from the deexcitation cascade following the projectile capture are detected. Since the γ rays cannot be resolved by the large crystal due to its long response time, they are all summed into the so-called sum peak. When the entry state populated by the captured projectile decays into the ground state, the energy of the sum peak can be expressed as

$$\gamma_\Sigma = E_{entry} - E_{g.s.} = E_{CM} + Q, \quad (1)$$

where the E_{entry} and $E_{g.s.}$ are the energy of the entry and ground state, respectively, E_{CM} is the projectile energy in the center-of-mass system and Q is the reaction Q-value.

The method has several advantages over the alternative technique (i.e., using the angular distribution of the emitted γ rays). Since the detector covers nearly 100% of the solid angle there is no need for angular distribution corrections, which greatly reduces the uncertainty. Additionally, the angular distribution method requires identification of all the γ rays feeding the ground state to account for all the possible branching paths of the decay of the entry states. Frequently, the γ-decay branchings are not known very well or the intensity of the γ rays is low and they cannot be resolved from the background, which may result in underestimation of the measured cross section. This is not the case with the summing technique, where each decay branch will be summed to the same energy (γ_Σ). The only peak structure in the spectrum that needs to be analyzed is the sum peak with the area proportional to the measured cross section.

3. Experimental procedure

The experiments were performed at the Nuclear Science Laboratory of the Univerity of Notre Dame using a γ-summing detector, HECTOR. The High EffiCiency TOtal absorption spectrometeR is an array of sixteen $4'' \times 8'' \times 8''$ NaI(Tl) crystals, each read out by two photomultipliers (see Fig. 2). Each segment of HECTOR is housed by a 1 mm aluminum casing and the crystals are assembled to form a 16-inch cube. A 60 mm bore hole through the

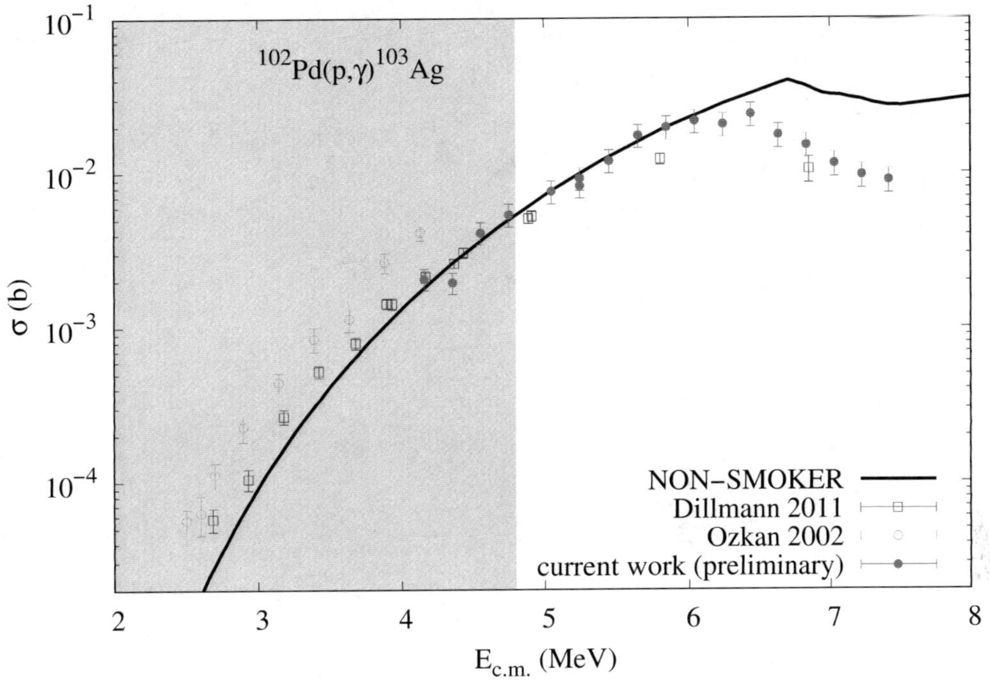

Figure 3. ^{102}Pd(p,γ)^{103}Ag cross section obtained with HECTOR (solid red circles) compared with previous results of Dillmann et al. [23] (open blue squares) and Ozkan 2002 [24] (open yellow circles). Solid black line denotes the cross section from the NON-SMOKER website [25]. The gray shaded area indicates the Gamow window.

array allows for placing the target in the center of the array without compromising the solid angle covered by the detector. Details of the detector can be found in [22].

Highly enriched targets of ^{102}Pd (78(1)% enriched) and ^{90}Zr (98(1)% enriched) were used during the experiment. The ^{102}Pd was bombarded with 1-10 nA proton beam, while a 10-25 nA beam of α particles was impinged on the ^{90}Zr target. In each case, the target was placed in the center of the detector to maximize the γ-summing efficiency.

The beam pipe within the setup was electrically insulated and served as a Faraday cup for measurement of the beam current. The charge deposited in the Faraday cup was collected by a charge integrator and recorded within the data acquisition system for continuous monitoring of the beam intensity. The energy range covered during the experiment was 4-7.5 MeV and 7.5-11.5 MeV for proton and α beam, respectively.

4. Results

The data analysis was performed in accordance with the method described in detail in [22]. For each beam energy, the integral under the sum-peak in the spectrum was obtained and the background from the incomplete summation was subtracted. The γ-summing efficiency was determined using a Geant4 simulation of the HECTOR array. The target thickness has been determined using the Rutherford backscattering technique. The uncertainty in the cross sections includes the statistical and systematic uncertainty of the sum-peak integral, 5% uncertainty in the target thickness and in the total beam current, and 10% uncertainty in the summing efficiency. The preliminary results are shown in Figures 3 and 4, for the ^{102}Pd(p,γ)^{103}Ag and

IOP Conf. Series: Journal of Physics: Conf. Series **1308** (2019) 012020 doi:10.1088/1742-6596/1308/1/012020

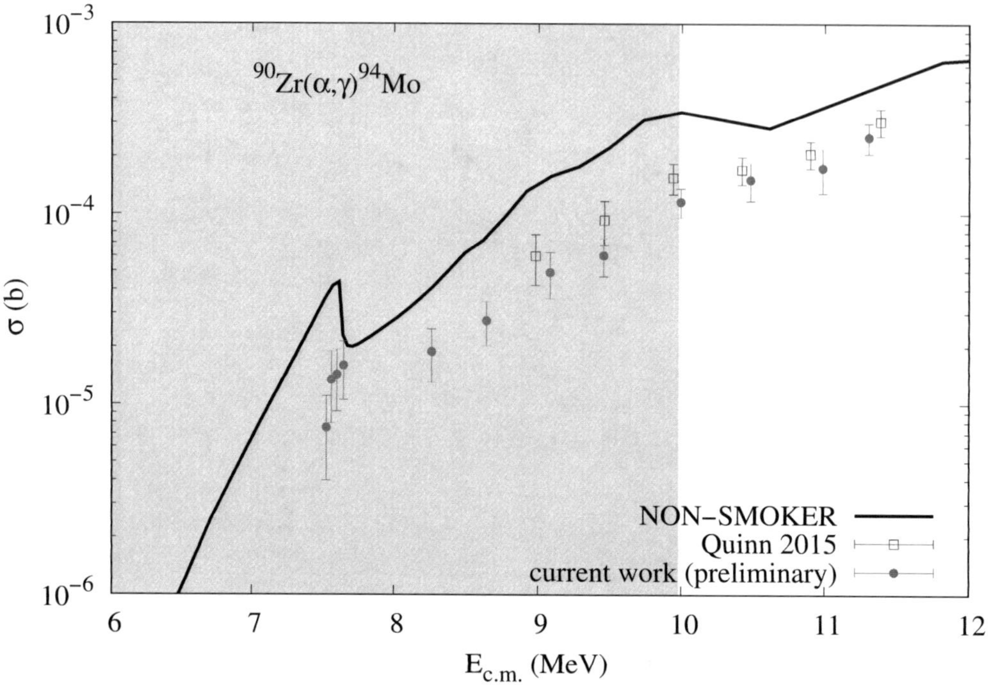

Figure 4. $^{90}Zr(\alpha,\gamma)^{94}Mo$ cross section obtained with HECTOR (solid red circles) compared with previous results of Quinn et al. [26] (open blue squares). Solid black line denotes the cross section from the NON-SMOKER website [25]. The gray shaded area indicates the Gamow window.

$^{90}Zr(\alpha,\gamma)^{94}Mo$ reactions, respectively.

In the case of $^{102}Pd(p,\gamma)^{103}Ag$ reaction, two previous measurements were found in the literature. The first one from Ozkan et al. [24] was performed using the activation technique, the second one by Dillmann et al. [23] utilized in-beam γ-ray detection using Ge detectors. There is a discrepancy of about a factor of two between the two data sets. The current measurement shows a very good agreement with the data from Dillmann et al. and extends the covered energy range towards higher energies. For comparison, the cross section obtained from NON-SMOKER is also shown in Figure 3. The model seems to under-predict the cross section at lower energies while it overestimates the capture cross section at energies above 6.5 MeV. In the energy range of 4-6.5 MeV a good agreement between the current work, data from Dillmann et al. and NON-SMOKER can be observed.

The $^{90}Zr(\alpha,\gamma)^{94}Mo$ reaction has previously been measured using the γ-summing technique by Quinn et al. [26]. Very good agreement between the previous measurement and current work can be seen in Figure 4. Additionally, the current measurement extends the experimentally covered energy range down to 7.5 MeV. NON-SMOKER overestimates the experimental cross sections by a factor of 2-3.

5. Conclusions

The first results from a γ-summing detector, HECTOR, at the University of Notre Dame were presented. Two measurements of capture reactions: $^{102}Pd(p,\gamma)^{103}Ag$ and $^{90}Zr(\alpha,\gamma)^{94}Mo$ were discussed and compared with the previous measurements found in the literature and with NON-

IOP Conf. Series: Journal of Physics: Conf. Series **1308** (2019) 012020 doi:10.1088/1742-6596/1308/1/012020

SMOKER calculations. In the case of the ^{102}Pd(p,γ)^{103}Ag reaction, a good agreement with the previous measurement of Dillmann et al. [23] was found. For the ^{90}Zr(α,γ)^{94}Mo reaction, the current results are in a very good agreement with the results of Quinn et al. [26] who utilized the same experimental technique. The results serve as a proof-of-principle for the newly commissioned detector.

6. Acknowledgments
This work was supported by NSF under grants No. PHY-1614442 (Simon), PHY-1713857 (Nuclear Science Laboratory), PHY-1430152 (JINA-CEE) and PHY-1102511 (NSCL).

References
[1] Burbidge E M, Burbidge G R, Fowler W A and Hoyle F 1957 *Rev. Mod. Phys.* **29** 547
[2] Clayton D, Fowler W, Hull T and Zimmerman B 1961 *Ann. Phys.* **12** 331
[3] Seeger P A, Fowler W A and Clayton D D 1965 *Astrophys. J. Supp. Ser.* **11** 121
[4] Cowley C R and Downs P L 1980 *Astrophys. J.* **236** 648
[5] Arnould M and Goriely S 2003 *Phys. Rep.* **384** 1
[6] Rauscher T 2011 *Proceedings of Science PoS(NIC XI)* 059
[7] Woosley S E and Howard W M 1978 *Astrophys. J. Suppl.* **36** 285
[8] Rauscher T, Heger A, Hoffman R D and Woosley S E 2002 *Astrophys. J.* **576** 323
[9] Rayet M, Arnould M, Hashimoto M, Prantzos N and Nomoto K 1995 *Astron. Astrophys.* **298** 517
[10] Woosley S E and Heger A 2007 *Phys. Rep.* **442** 269
[11] Howard W M, Meyer B S and Woosley S E 1991 *Astrophys. J. Lett.* **373** L5
[12] Howard W M and Meyer B S 1993 *Proc. 2nd Int. Conf. Nuclei in the Cosmos* 575
[13] Kusakabe M, Iwamoto N and Nomoto K 2005 *Nucl. Phys.* **A758** 459
[14] Kusakabe M, Iwamoto N and Nomoto K 2011 *Astrophys. J.* **726** 25
[15] Travaglio C, Röpke F K, Gallino R and Hillebrandt W 2011 *Astrophys. J.* **739** 93
[16] Hauser W and Feshbach H 1952 *Phys. Rev.* **87** 366
[17] Rauscher T and Thielemann F K 2000 *Atomic Data and Nuclear Data Tables* **75** 1
[18] Tsagari P, Kokkoris M, Skreti E, Karydas A G, Harissopulos S, Paradellis T and Demetriou P 2004 *Phys. Rev. C* **70** 015802
[19] Harissopulos S, Lagoyannis A, Spyrou A, Zarkadas C, Galanopoulos S, Perdikakis G, Becker H W, Rolfs C, Strieder F, Kunz R, Fey M, Hammer J W, Dewald A, Zell K O, von Brentano P, Julin R and Demetriou P 2005 *J. Phys. G* **31** S1417
[20] Spyrou A, Becker H W, Lagoyannis A, Harissopulos S and Rolfs C 2007 *Phys. Rev. C* **76** 015802
[21] Simon A, Quinn S, Spyrou A, Battaglia A, Beskin I, Best A, Bucher B, Couder M, DeYoung P, Fang X, Görres J, Kontos A, Li Q, Liddick S, Long A, Lyons S, Padmanabhan K, Peace J, Roberts A, Robertson D, Smith K, Smith M, Stech E, Stefanek B, Tan W, Tang X and Wiescher M 2013 *Nucl. Instr. Meth. A* **703** 16
[22] Reingold C S, Olivas-Gomez O, Simon A, Arroyo J, Chamberlain M, Wurzer J, Spyrou A, Naqvi F, Dombos A C, Palmisano A, Anderson T, Clark A M, Frentz B, Hall M R, Henderson S L, Moylan S, Robertson D, Skulski M, Stech E, Strauss S Y, Tan W P and Vande Kolk B 2019 *The European Physical Journal A* **55** 77
[23] Dillmann I, Coquard L, Domingo-Pardo C, Kappeler F, Marganiec J, Uberseder E, Giesen U, Heiske A, Feinberg G, Hentschel D, Hilpp S, Leiste H, Rauscher T and Thielemann F K 2011 *Phys. Rev. C* **84** 015802
[24] Özkan N, Murphy A, Boyd R, Cole A, Famiano M, Güray R, Howard M, Sahin L, Zach J, deHaan R, Görres J, Wiescher M, Islam M and Rauscher T 2002 *Nuclear Physics A* **710** 469
[25] Raucher T and Thielemann F K 2001 *Atomic Data and Nuclear Data Tables* **79** 47
[26] Quinn S J, Spyrou A, Simon A, Battaglia A, Bowers M, Bucher B, Casarella C, Couder M, DeYoung P A, Dombos A C, Görres J, Kontos A, Li Q, Long A, Moran M, Paul N, Pereira J, Robertson D, Smith K, Smith M K, Stech E, Talwar R, Tan W P and Wiescher M 2015 *Phys. Rev. C* **92** 045805

IOP Conf. Series: Journal of Physics: Conf. Series **1308** (2019) 012021 doi:10.1088/1742-6596/1308/1/012021

Theoretical approaches to the 3α break-up of ^{12}C

Robin Smith[1†], Jack Bishop[2], Carl Wheldon[3], Tzany Kokalova[3]
[1]Faculty of Science, Technology & Arts, Sheffield Hallam University, Sheffield, S1 1WB, UK
[2]Department of Physics & Astronomy and Cyclotron Institute, Texas A&M University, College Station, TX 77843, USA
[3]School of Physics & Astronomy, University of Birmingham, Edgbaston, Birmingham, B15 2TT, UK

E-mail: †robin.smith@shu.ac.uk

Abstract. Two recent experiments have indicated that the break-up of the ^{12}C Hoyle state is dominated by the sequential ^{8}Be(g.s.) + α decay channel. The rare direct 3α decay was found to contribute with a branching ratio of less than 0.047% (95% C.L.). However, the ability of experimentalists to successfully disentangle these two competing decay channels relies on accurate theoretical predictions of how they each manifest in phase space distribution of the three break-up α-particles. The following paper reviews the current theoretical approaches to calculating the break-up of the Hoyle state and introduces a semi-classical WKB approach, which adequately reproduces the results of more sophisticated calculations. It is proposed that a more accurate upper limit on this branching ratio may be obtained if these new theoretical results are taken into account when analysing experimental data.

1. Introduction

There has recently been a renewed interest in studying the decay modes of the Hoyle state in ^{12}C [1, 2]. In astrophysics, the Hoyle state is crucial for the synthesis of carbon in the helium burning phase of main sequence stars. The presence of a 0^+ resonance near to the 3α threshold was first predicted by Fred Hoyle to boost the triple-α capture process by seven orders of magnitude [3]. The prediction was successfully verified experimentally and this famous state bears Hoyle's name. Due to the way the Hoyle state resonance is synthesised in nature, it was proposed that this state is an excellent candidate for the presence of α-clustering, a feature of nuclei that has been discussed throughout the history of nuclear physics [4]. If we assume the ^{12}C nucleus to be made up of three α-particles, it leads to the question of what configuration they take within the nucleus.

The U(7) algebraic cluster model (ACM), which assumes an equilateral triangle 3α structure with $\mathcal{D}_{3h}$ point symmetry, well reproduces the low energy portion of the ^{12}C excitation spectrum [5]. In this model, the ground state is a compact 3α structure and the Hoyle state has a larger average separation of the α-clusters, as illustrated in figure 1. Contrastingly, an *ab initio* lattice approach using chiral effective field theory has predicted the state to have a 'bent-arm' configuration. Furthermore, other *ab initio* fermionic molecular dynamics (FMD) and antisymmetrized molecular dynamics (AMD) calculations predict a clear α + ^{8}Be configuration in the intrinsic frame [6]. It has been suggested that the energy distribution of the three α-particles emitted during the decay of ^{12}C may provide some indication of the initial configuration of those α-particles before the decay, meaning that precise break-up measurements could be a powerful structural probe.

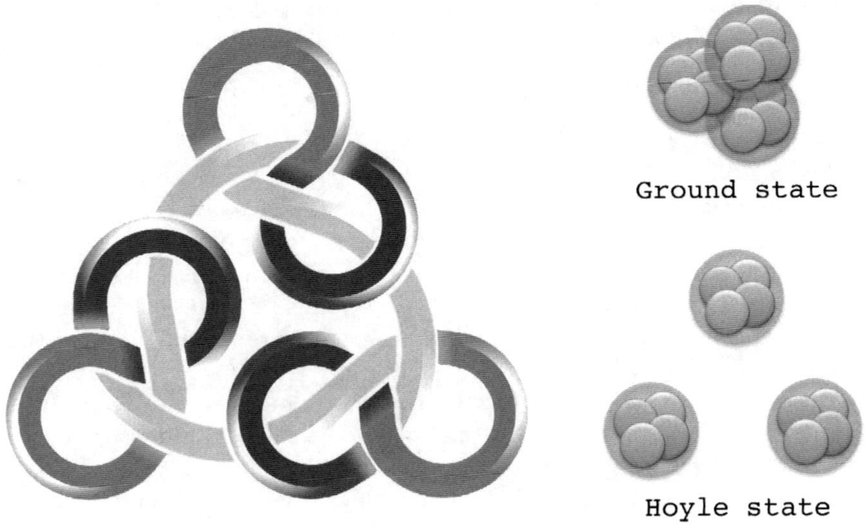

Figure 1. (Colour online) Traditional Celtic knot design illustrating the 3α structures in ^{12}C. The dark (blue) loops represent the compact ground state and the lighter (red) loops represent the excited Hoyle state, which possesses a large radius. The two structures are linked by a single knot, alluding to the fact that these two manifestations are both part of the same nucleus.

One particular theory for the structure arises from the idea that the bosonic nature of the α-particle could dominate the dynamics of this nuclear state. In agreement with the predictions of the ACM, the charge form factor for inelastic electron excitation from the ground state to the Hoyle state indicates that the Hoyle state has a large radius and hence a larger average separation of the alpha clusters [7]. This feature is illustrated in figure 1. Under this assumption, the important degrees of freedom could then be considered bosonic rather than fermionic, producing the nuclear analogue of an atomic Bose Einstein Condensation (BEC) [8]. Under these conditions there should be no structural preference for either a sequential, ^{8}Be + α, or a direct 3α decay and so the branching ratio should depend only on the available phase space and penetrabilities through the decay barrier [9]. The present experimental upper limits on the direct decay mode cannot exclude this possibility due to theoretical uncertainties [10], so more experimental work is needed to further constrain the direct decay branching ratio.

Until recently, when comparing experimental decay data to theoretical predictions, simple phenomenological models for sequential and direct 3α break-up were used. For the sequential case, a narrow intermediate ^{8}Be ground state resonance at 92 keV was modelled. However, the possible enhancement of this resonance at higher excitations in the form of its *ghost* was not considered in detail. Further, the direct decay mode has historically been modelled purely phenomenologically. Uniform decays to the available phase space, equal α-particle energies, and collinear cases have been compared with data, but none of these have any firm theoretical grounding.

One of the most recent studies made a more accurate theoretical description of the direct decay phase space distribution using the WKB method in hyperspherical coordinates. Using this model, an improved upper limit of 0.035% was placed on the direct decay branching ratio [10].

IOP Conf. Series: Journal of Physics: Conf. Series **1308** (2019) 012021 doi:10.1088/1742-6596/1308/1/012021

A description of this theoretical methodology and the PeTA code used in the calculations is the focus on this paper. Later, other theoretical approaches to describing the direct decay are discussed and compared with the semi-classical approximation. These calculations indicate that some of the data in experiments that are labelled as *direct* decay, may in fact be a different manifestation of the *sequential* decay. It is argued that a reanalysis of the latest experimental data is needed.

2. A semi-classical WKB approach to 3α decay

2.1. Sequential decay

The semi-classical approach presented in reference [11] was extended in order to examine the energy distribution of α-particles emitted during the direct decay of the Hoyle state. The present section builds on the work of reference [10]. The WKB approximation is a method for obtaining an approximate solution to the time-independent Schrödinger equation in one-dimension. In this case, it is used to calculate 3-body tunnelling rates through Coulomb potential barriers.

In the simple two-body α + ^{8}Be decay of ^{12}C, the Schrödinger equation can be easily solved for the relative motion of the two daughter nuclei and the Coulomb interaction between them. For a general 2-body decay, where the two daughters have charges Z_1 and Z_2, a separation of r, and a relative angular momentum of ℓ, the potential is given as the sum of the Coulomb and centrifugal terms by the equation

$$V_\ell(r) = \frac{Z_1 Z_2 e^2}{4\pi\epsilon_0 r} + \frac{\ell(\ell+1)}{2\mu r^2}, \tag{1}$$

where μ is the reduced mass of the two-particle system. The potential for the $\ell = 0$ α + ^{8}Be decay of the Hoyle state is shown in the left panel of figure 2. Using this potential, the tunnelling penetrability "T" may then be calculated using the WKB method as

$$T = \frac{1}{1+e^{2S}} \approx e^{-2S}, \quad \text{where} \quad S = \frac{1}{\hbar}\int_{r_0}^{r_1} dr\sqrt{2\mu\left(V_\ell(r) - E\right)}, \tag{2}$$

E is the energy of the system above the two-body decay threshold, r_0 is the matching radius and r_1 is the classical turning point. This method works well and calculates a penetrability for the α + ^{8}Be decay of the Hoyle state within 5% of the exact value calculated using the regular and irregular Coulomb wave functions evaluated at the channel radius [12].

2.2. Direct decay

Unlike in the simple sequential decay case, a direct decay into three α-particles permits a variety of different configurations. These vary from an extreme collinear decay, where two of the α-particles are emitted back-to-back, leaving the third stationary, all the way through to an equal energies case, where the α-particles are emitted at 120° to one another. These cases are pictured in the inset of the right panel of figure 2. Importantly, these different decay configurations possess different Coulomb barriers which are also shown in the right panel of figure 2. Therefore, even though a direct decay mechanism may uniformly sample the phase space for a given decay, certain 3α configurations are preferred, and these Coulomb effects will heavily influence the phase space distribution in a direct decay.

To model this kind of decay, we parameterise the system using a quantity called the hyperradius, ρ, which depends on the masses of each particle m_i and their separation r_{ik}. It is defined as

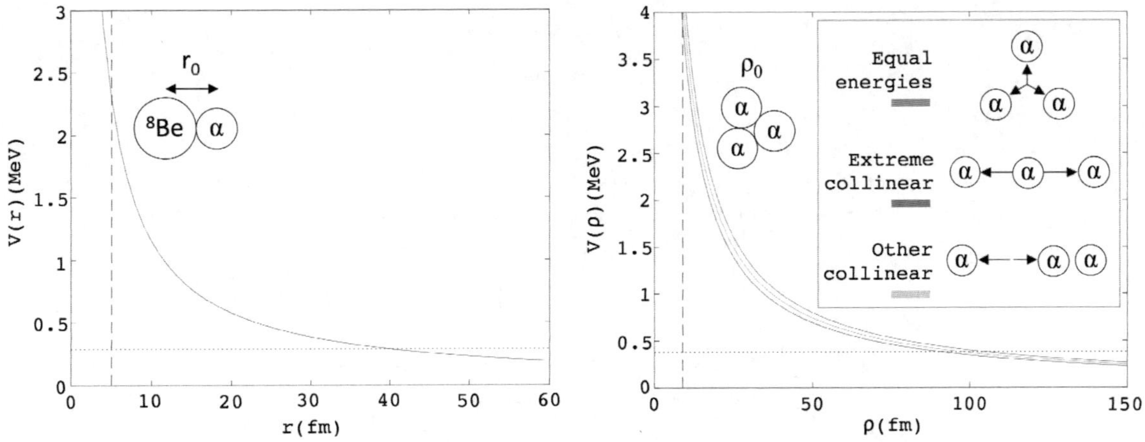

Figure 2. (Colour online) Left panel: the Coulomb potential for the α + ^{8}Be decay of the Hoyle state. Right panel: the Coulomb potential as a function of the hyperradius for various direct 3α break-up configurations of the Hoyle state.

$$\rho^2 \equiv \frac{1}{mM} \sum_{i<k}^{3} m_i m_k r_{ik}^2, \tag{3}$$

where $M = \sum_j^3 m_j$ and m is an arbitrary normalisation mass. If the particles are emitted during the decay with certain energies (velocities), as the decay progresses, the separation between the particles, and hence, the hyperradius of the system, will increase linearly. This only applies under the strict assumption that we ignore final-state Coulomb interactions. In reality, the Coulomb interactions between the particles during the decay will perturb their paths and hence the Coulomb barrier. Here, we assume the particles follow their initial trajectory, unperturbed, and hence tunnel through a constant potential barrier. The validity of this assumption is addressed in the approach presented in section 3.

A particular decay orientation is defined by positive scaling constants s_{ik}, which are a constant of the decay, and are calculated by normalising the relative distances between the particles by the hyperradius as

$$s_{ik}^2 \equiv \frac{r_{ik}^2}{\rho^2}. \tag{4}$$

The Coulomb potential, accounting for the interactions between all α-particles can be parameterised by the hyperradius and scaling constants as

$$V_C(\rho) = \sum_{i<k}^{3} \frac{Z_i Z_k e^2}{r_{ik}} = \frac{1}{\rho} \sum_{i<k}^{3} \frac{Z_i Z_k e^2}{s_{ik}}. \tag{5}$$

Although it is not needed for the decay of the 0^+ Hoyle state, a general 3-body decay must account for the centrifugal barrier term

$$V_K(\rho) = \frac{\hbar(K+3/2)(K+5/2)}{2m\rho^2}, \tag{6}$$

where m is the normalisation mass and K is the hypermomentum quantum number. In this model it is assumed that only the lowest value partial wave contributes because it has the lowest barrier [11, 13]. In general, the total potential barrier, $V(\rho)$, is the sum of equations 5 and 6. This may be used in conjunction with equation 2 to give the tunnelling penetrability as

$$T = \frac{1}{1+e^{2S}} \approx e^{-2S}, \quad \text{where} \quad S = \frac{1}{\hbar}\int_{\rho_0}^{\rho_1} d\rho\sqrt{2m\,(V(\rho)-E)}, \tag{7}$$

where ρ_0 corresponds to the hyperradius where all three α-particles are touching, shown in the right panel of figure 2 and ρ_1 is the classical turning point. The `PeTA` code utilises this result.

2.3. Penetrability of Three Alphas (`PeTA`) code and results

The Penetrability of Three Alphas (`PeTA`) code was used to determine a realistic distribution of the three α-particle energies during the direct decay of the Hoyle state. The `Matlab` source code may be obtained from reference [14]. It is a Monte-Carlo code which uniformly samples the available phase space for a given direct decay and generates the permitted non-relativistic 3α momenta subject to the constraints of energy and momentum conservation. To quantify the total phase space of an n-body decay, all of the kinematically-allowed configurations must be integrated over. The total phase space for an n-body decay into particles of mass m_i, with an energy over threshold of E, is given by the equation

$$W(E,n) = \left(\frac{1}{\sum_{b=1}^{n} m_b}\prod_{b=1}^{n} m_b\right)^{3/2} \frac{(2\pi)^{3(n-1)/2}}{\Gamma(3(n-1)/2)} E^{3n/2-5/2}, \tag{8}$$

where Γ is the Gamma function. The full derivation of this function is shown in reference [10]. From this alone, it is possible to estimate the direct 3α decay branching ratio as $W(E_3,3)/W(E_2,2) = 1.8 \times 10^{-3} = 0.18\%$.

The `PeTA` code generates a user-defined number of 3-body direct decay events within the allowed phase space (chosen here for illustration to be 10^6). For each event, the unique orientations of the α-particles result in a different Coulomb barrier, and the resulting penetrability is calculated using equation 7. The integral is calculated numerically using the trapezium rule. The decay amplitude for a particular orientation is weighted by this penetrability. A plot of the simulated events is shown in the right panel of figure 3 in the form of a Dalitz plot [15]. In contrast, uniform direct decays without any penetrability factor yield a uniform coverage of the Dalitz plot. For comparison, the sequential decay, where one of the α-particles carries away around half of the decay energy, is shown in the left panel of figure 3. This plot includes the effects of the experimental resolution. The coordinates of the Dalitz plots are written in terms of the fractional α-particle energies in the frame of the decaying ^{12}C, ϵ_i. The coordinates are defined as $x = \sqrt{3}(\epsilon_1 - \epsilon_3)$ and $y = 2\epsilon_2 - \epsilon_1 - \epsilon_3$.

Typically, when analysing experimental data, a uniform coverage of the Dalitz plot is used to model the direct decay. The more realistic direct decay profile peaks towards the centre of the Dalitz plot and has less overlap with sequential decays, which appear as a straight band. When considering this type of direct decay, named DDP^2 (Direct Decay + Phase space + Penetrability), an improved upper limit of the direct decay can be placed at 0.035%, compared with 0.047% when using the standard direct decay profile [10].

IOP Conf. Series: Journal of Physics: Conf. Series **1308** (2019) 012021 doi:10.1088/1742-6596/1308/1/012021

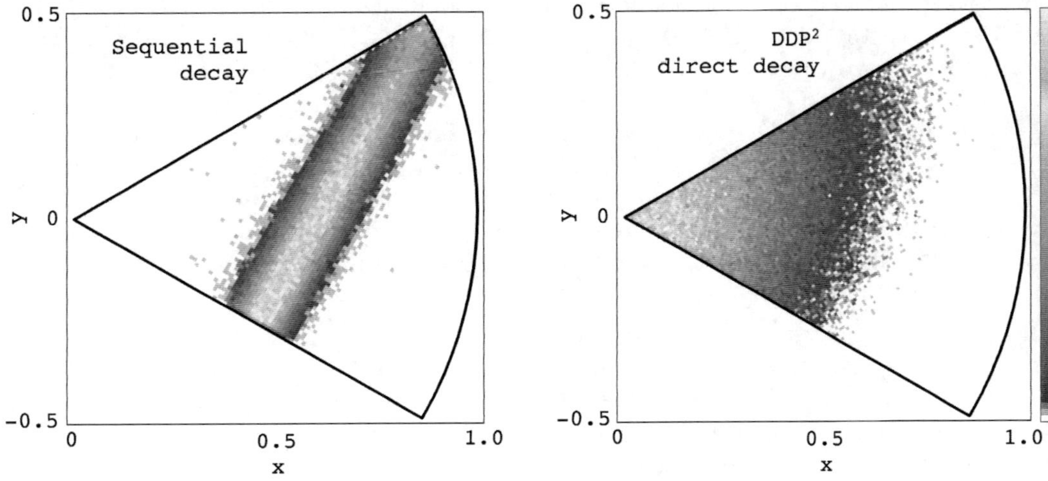

Figure 3. (Colour online) Left: Dalitz plot for sequential decays (including experimental resolution). Right: Weighted phase space distribution including the effects of the WKB 3-body penetrability (excluding experimental resolution). The coordinates are defined in the main text.

3. An R-matrix approach

In 2018, Refsgaard *et al.* developed a sequential R-matrix model to describe the Hoyle state break-up [15]. They noted that even in the α + ^{8}Be sequential decay, some features that may resemble a direct 3α decay naturally appear in the phase space distribution of the α-particles. This arises due to an unusual feature of near-threshold resonances called the *ghost anomaly* [16]. The R-matrix line shape for an isolated single channel resonance is given by the equation

$$w(E) = \frac{1}{\pi} \frac{\Gamma_i/2}{[E_i - E - \gamma^2(S - B)]^2 + \Gamma_i^2/4}, \quad (9)$$

where S and B are the shift function and boundary condition for the channel. Since the channel width Γ_i is energy dependent ($\Gamma_i(E) = 2P_\ell(E)\gamma_i$ where γ_i is the reduced channel width), as the energy above threshold increases, so does the decay penetrability through the Coulomb barrier, meaning that $\Gamma_i(E)$ also increases. This modifies the standard Breit-Wigner lineshape of the resonance and permits an extra yield above the main resonance peak. It is this extra yield that may be mistaken for direct decays in experimental data.

The theoretical intensity of the ghost anomaly is highly dependent on the channel radius; a free parameter in R-matrix theory. Using this method with sensible channel radii, the ghost anomaly is expected to appear with a strength ranging between 0.01% and 1%. This is approaching the levels of sensitivity seen in the latest experiments and hence must be factored in when attempting to evaluate any direct decay contribution in the data.

Another important feature of this R-matrix approach is to include a final-state Coulomb interaction (FSCI) between the three α-particles. The secondary break-up of ^{8}Be happens within the Coulomb field of the initial emitted α-particle. This is built into the model by replacing the penetrabilities of the α + ^{8}Be pair by the product of penetrabilities for the $\alpha + \alpha_1$ and $\alpha + \alpha_2$ pairs (where α_1 and α_2 are the two α-particles resulting from the ^{8}Be decay) once the α

IOP Conf. Series: Journal of Physics: Conf. Series **1308** (2019) 012021 doi:10.1088/1742-6596/1308/1/012021

+ ^{8}Be pair have separated to some distance. This model nicely fits experimental data from the break-up of the 1^+ state at 12.71 MeV and it would be desirable to test this model against the latest high precision Hoyle state decay data [1, 2].

This approach models direct decay by approximating this as a sequential decay through a very broad resonance at high excitation. The results were seen to be quite insensitive to the energy and width of this resonance. The phase space distribution predicted by this model produces a Dalitz plot (right panel of figure 3 in reference [15]) very similar to that shown by the right panel of our figure 3, adding credibility to the semi-classical WKB approximation.

4. Full 3-body quantum mechanical calculations

Although Coulomb interactions between the α-particles modify their final state configuration, the decay barriers are significantly altered when including an attractive $\alpha-\alpha$ interaction. Hence, several attempts have been made to understand the decay and structure of the Hoyle state by performing full three-body quantum mechanical calculations.

Alvarez-Rodríguez *et al.*, calculated the 3α decays of states in ^{12}C using the hyperspherical adiabatic expansion method of the Faddeev equations [17]. The interaction term was chosen to vary only with the hyprerradius. This approach simultaneously describes all the possible interactions in a system of three particles in a fully quantum mechanical way. The angular part of the Schrödinger equation is solved and the full wave function is then expanded using a basis of angular wave functions [13]. This allows these components of the wave function to be examined at fixed hyperradii, as the system separates. It was found that the Hoyle state is dominated by the $n = 1$ angular basis function over all values of the hyperradius from $\rho = 0$ to 80 fm. This indicates that the angular wave function does not significantly change during the decay and that the final state distribution of α-particle energies may be a good reflection of the initial structure. The calculations also predict a direct decay branch of the Hoyle state at 1%; significantly larger than the best experimental upper limits.

The method also permits the energy distributions of the final state α-particles to be calculated, but the results for the Hoyle state are not discussed in the work of Alvarez-Rodríguez *et al.* However, a theoretical study by Ishikawa predicted the direct 3α component of the decay of the Hoyle state [18] and at the same time calculated the 3α phase space distribution. In that work, the wave function for the reaction γ + ^{12}C $\rightarrow 3\alpha$ was solved using the Faddeev three-body formalism. The α-particles were treated as bosons and their underlying fermionic structures were considered to be incorporated into the various $\alpha - \alpha$ interactions used.

The phase space distribution of the three α-particles during the decay [18] is shown as the right panel of figure 4. This is in remarkably good agreement with the R-matrix model of Refsgaard *et al.* [15] shown as the left panel of figure 4. The majority of events that may be considered as *direct decay* lie near to the sequential decay band and so could just be a manifestation of the ^{8}Be ghost anomaly. The regions of the Dalitz plot predicted by Ishikawa were split up as in figure 1 b) in reference [18], in order to approximate the "direct decay" branching ratio. This indicates that the direct decay should contribute at the total level of $\approx 0.1\%$, with DDE (equal energies) and DDL (collinear) direct decay contributions at the levels of 0.005% and 0.03% respectively.

5. Summary

There are numerous ways to examine the 3α decay of the Hoyle state; from the semi-classical WKB approximation through to a full three-body quantum mechanical approach. Certain predictions of each of these models overlap. It was seen that the phase space distributions

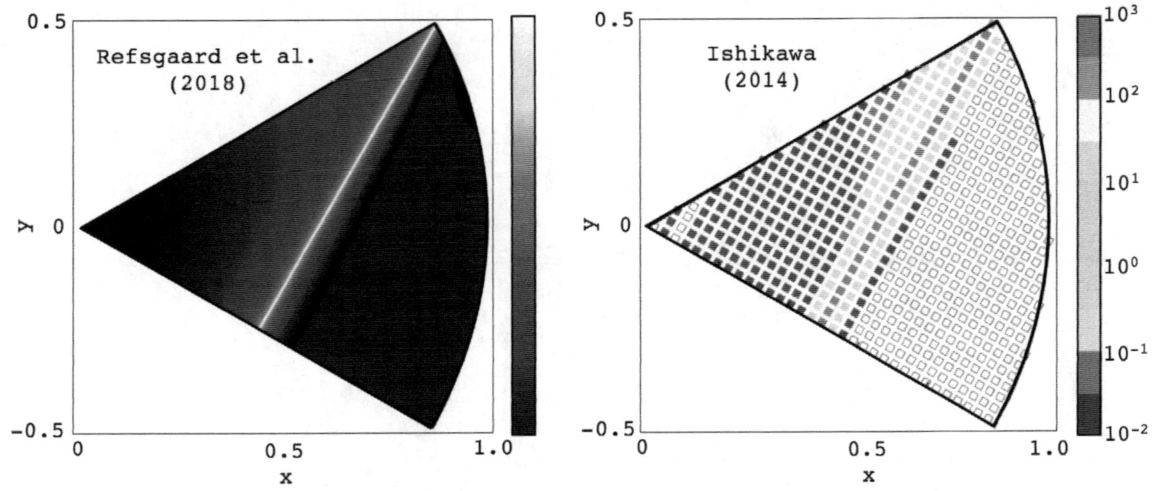

Figure 4. (Colour online) Theoretical Dalitz plot distributions by Refsgaard *et al.* [15] (left panel) and Ishikawa [18] (right panel). Both are plotted with a logarithmic colour scale.

for direct decay predicted by the WKB and R-matrix methods are very alike. Similarly, the distribution calculated using the Faddeev three-body formalism reproduces the main features of the R-matrix sequential model. The latest Hoyle state decay data need to be reanalysed and compared with these physically motivated α energy distributions, rather than the phenomenological models typically used.

6. Acknowledgements

Discussions with J. Refsgaard are gratefully acknowledged.

References

[1] Smith R, Kokalova Tz, Wheldon C, Bishop J, Freer M, Curtis N and Parker D J 2017 *Phys. Rev. Lett.*, **119** 132502.
[2] Dell'Aquila D, Lombardo I, Verde G, Vigilante M, Acosta L, Agodi C, Cappuzzello F, Carbone D, Cavallaro M, Cherubini S, Cvetinovic A *Phys. Rev. Lett.* **119** 132501.
[3] Hoyle F, Dunbar D N F, Wenzel W A, and Whaling W 1953 *Phys. Rev.* **92** 1095.
[4] Freer M 2007. *Rep. Prog. Phys.* **70** 2149.
[5] Marín-Lámbarri D J, Bijker R, Freer M, Gai M, Kokalova Tz, Parker D J and Wheldon C 2014 *Phys. Rev. Lett.* **113** 012502.
[6] Freer M, and Fynbo H O U 2014 *Prog. Part. Nucl. Phys.* **78**.
[7] Danilov A, Belyaeva T, Demyanova A, Goncharov S and Ogloblin A 2009 *Phys. Rev.* C **80** 054603.
[8] Tohsaki A, Horiuchi H, Schuck P and Röpke G 2001 *Phys. Rev. Lett.* **87** 192501.
[9] Kokalova Tz, Itagaki N, von Oertzen W and Wheldon C 2006 *Phys. Rev. Lett.* **96** 192502.
[10] Smith R 2017 Ph.D. Thesis, University of Birmingham.
[11] Garrido E, Fedorov D V, Jensen A S and Fynbo H O U 2005 *Nucl. Phys.* A **748** 27.
[12] Wheldon, C `ckin` 2-body kinematics code $http://www.np.ph.bham.ac.uk/research_resources/programs/$ accessed 24-Sep-2014.
[13] Nielsen E, Fedorov D V, Jensen A S and Garrido E 2001 *Phys. Rep.* **347** 373.
[14] Smith R `PeTA` version 1.0, Matlab File Exchange, 70496, accessed: 06-March-2019.
[15] Refsgaard J, Fynbo H O U, Kirsebom O S, Riisager K 2018 *Phys. Lett.* B **779** 414.
[16] Barker F C, Treacy P B 1962 *Nucl. Phys.* **38** 33.
[17] Alvarez-Rodríguez R, Jensen A S, Fedorov D V, Fynbo H O U, Garrido E 2007 *Phys. Rev. Lett.* **99** 072503.
[18] Ishikawa S 2014 *Phys. Rev.* C **90** 061604.

IOP Conf. Series: Journal of Physics: Conf. Series **1308** (2019) 012022 doi:10.1088/1742-6596/1308/1/012022

Neutron-induced reactions investigated via the Trojan Horse Method

R Spartá[1,2], C Spitaleri[1,2], L Lamia[1,2], G L Guardo[2], S Cherubini[1,2], M Gulino[1,3], M La Cognata[2], R G Pizzone[2], G G Rapisarda [2], S Romano[1,2], M L Sergi[2], A Tumino[1,3]

[1] Dipartimento di Fisica e Astronomia E. Majorana - Universitá degli Studi di Catania, Catania, Italy
[2] Laboratori Nazionali del Sud - INFN, Catania, Italy
[3] Universitá KORE, Enna, Italy

E-mail: rsparta@lns.infn.it

Abstract. The Trojan Horse Method has been applied to many neutron-induced reactions using the deuteron as a virtual source of neutrons, to explore wide energy regions of interest for astrophysics and applied physics and to investigate the suppression of the centrifugal barrier, that is one of the key advantages of this Method. The neutron-induced experimental campaign has already concerned ^{17}O(n,α)^{14}C, ^{6}Li(n,α)^{3}H, ^{7}Be(n,p)^{7}Li, ^{7}Be(n,α)^{4}He, ^{18}F(n,α)^{15}O and ^{14}N(n,p)^{14}C, and very recently ^{25}Mg(n,α)^{22}Ne and ^{27}Al(n,p)^{27}Mg, while others are planned to be measured soon, thus influencing different astrophysical scenarios. Particular attention is dedicated to a new measurement regarding the ^{10}B(n,α)^{7}Li reaction, aimed to disentangle the ^{7}Li ground state contribution from its first excited state to the cross section.

1. Introduction

Dealing with neutron induced reactions measurements is challenging and complex as much as they turn to be always more necessary. Areas of interest of their applications range from pure nuclear physics (nuclear structure and dynamics) to nuclear astrophysics, where n-induced reactions are fundamental basically in any astrophysical scenario. The same holds for applied physics, where these reactions are nowadays used to study and understand our cultural heritage (as an example the european Ancient Charm project [1]). Some of them, as ^{10}B(n,α)^{7}Li, are also used in nuclear energy power plants (to monitor the neutron fluxes in the plant), and in medicine, as it is discussed in section 3.2.

2. The need of indirect methods

Measuring neutron induced reactions is anyhow challenging. It is not possible to get a pure target of neutrons and the beam is not easy to have. Indeed, a primary neutron producer reaction, such as ^{7}Li(p,n)^{7}Be, is necessary with low energy accelerators, but this will not produce monoenergetical neutrons. Moreover, it is not possible to accelerate neutrons in the usual ionization way. Despite many attempts are done in facilities spread around the world, it seems impossible to get an intense monoenergetical neutron beam.

IOP Conf. Series: Journal of Physics: Conf. Series **1308** (2019) 012022 doi:10.1088/1742-6596/1308/1/012022

Also the way to obtain the cross sections with the detailed balance principle applied to the inverse reactions is problematic, because of the low efficiency problems connected with neutron detection.

For all these reasons, the use of indirect methods is required to achieve results useful for pure and applied physics.

2.1. The Trojan Horse approach

The Trojan Horse Method (THM) [2] [3] results to be the best indirect technique for this case. It was born to determine bare nucleus S-factor at astrophysical energies, because it allows to obtain the cross section without the strong suppression at very low energies due to the Coulomb barrier.

This is made selecting the quasi free mechanism by the cross section of an appropriate 3-body reaction measured in the lab, where one of the two interacting particles is called the Trojan Horse nucleus, and has a very high probability of clusterization in two particles. In the peculiar circumstances of this mechanism, one of them will partecipate and the other will act as a spectator to the binary reaction, that is the goal of the experiment.

In recent years, THM has also been applied to neutron-involving reactions: for the case of neutron outgoing, it avoids the use of neutron detectors, reconstructing its kinematics by the other two outcoming particles of the reaction: this has been very helpful for the cases of ^{2}H(d,n)^{3}He [4] and ^{13}C(α,n)^{16}O [5], where their cross section have been extracted detecting the charged outcoming particle and the spectator proton and deuteron, respectively. Instead, for the case when neutron is inducing the reaction, using the very well known p+n structure of deuteron, where their intercluster motion in $l = 0$ is well known and described by the Hultén function.

Using a CD_2 target (plastic and easy to get) has revealed to be a very cheap and simple way to have a perfect virtual neutrons source. Moreover, the use of this indirect method let us need only one beam energy to get a wide energy range for the excitation function.

3. Reactions measured

The first reaction used to test the Method in the neutron circumstances has been the ^{6}Li(n,α)^{3}H, whose 2-body cross section has been extracted by the ^{6}Li(d,αp)^{3}H measured at Laboratori Nazionali del Sud (LNS) in Italy. This marked the first evidence that deuteron could be rightly used as neutron virtual source and was the first direct insight of the Trojan Horse cross section behaviour about the centrifugal barrier effects: it is evident from data that TH cross section is in agreement with direct results only if it is multiplied for the on-shell centrifugal effects [6] [7].

Another neutron induced reaction stuedied with THM is ^{17}O(n,α)^{14}C, that turns out to be of great interest for the astrophysical scenario of s-process nucleosynthesis, which is all about neutron captures [8]. In that context, reactions as ^{17}O(n,α)^{14}C are called neutron-poisons, because the total amount of neutron available for s-process depends on their reaction rate. With THM a new measurement has been performed in the energy range of astrophysical interest, resulting in a reaction rate quite different from the ones found in literature, and astrophysical consequences have still to be evaluated [9].

Partial results of a preliminary test to get the cross section of ^{10}B(n,α)^{7}Li have been already published [10], while the final ones are going to be subimitted for publication. However, because of the particular importance of this reaction for applications, a new and very precise measurement has been performed and is still under analysis, as reported in section 3.2.

3.1. Results under analysis and applications to exotic beams

Great efforts have been put on the measurement of ^{7}Be(n,α)^{4}He and ^{7}Be(n,p)^{7}Li, made in two runs (the first in LNL, Italy and the second at CRIB, Japan) to get these two cross sections

IOP Conf. Series: Journal of Physics: Conf. Series **1308** (2019) 012022 doi:10.1088/1742-6596/1308/1/012022

very relevant for the Big Bang Nucleosynthesis scenario. Good results coming out from these experiments [11] [12] are a proof that TH perfectly suits the needs of measurements involving exotic beams, such as ^{7}Be.

Also ^{14}N(n,p)^{14}C has been measured at LNS, because it is the first n-poison of s-process, mentioned above. Results are still under analysis [13].

Moreover, in the very recent past other two neutron induced reactions have been measured, both at LNS. The first was the ^{25}Mg(n,α)^{22}Ne to get the ^{22}Ne(α,n)^{25}Mg cross section using the detailed balance principle. The latter is, indeed, one of the most influencing reaction for the star nucleosynthsis scenario, which hardly impacts the s-process and all the rest of the life of a star.

The second one is ^{27}Al(n,p)^{27}Mg, that has been the first step of a long term campaign of measurement involving Al isotopes, which will very soon involve also exotic beams. This has been thought to give contraints to reaction rates involving the abundance of ^{26}Al, very important to trace the active nucleosynthesis in the Galaxy.

3.2. The case of $^{10}B(n,\alpha)^{7}Li$

This peculiar neutron induced reaction is of interest for the nuclear energy production in power plants, to have control on reactions induced on the percentage of ^{10}B in the natual boron, considering that ^{11}B(p,$\alpha\alpha$)^{4}He is the only good candidate for a clean (aneutronic) energy obtained by fusion. Moreover, it is very important for medicine, considering that the cure protocol of eye melanomas and glioblastoma mulltiforme [14] and the synovial ablation for rheumatoid artritis [15] consist in inducing this reaction inside the patient. For purposes of medicine, an energy range from 0 to 100 keV is of interest in the excitation function, while in the energy production case from 0 to 15 MeV. However, cross section measurement available in literature are still not exhaustive for these applications.

This reaction has already been measured using Trojan Horse Method [10], but a new run has been performed in order to separate the two main contributions to this cross section, coming from the ground state and from the first excited state of the ^{7}Li, separated by 0.477 MeV.

The experimental run has been performed in 2014 in Laboratori Nazionali del Sud (LNS, Catania), using a 28 MeV ^{10}B beam impinging on a CD_2 target, measuring ^{10}B(d,α ^{7}Li)H cross section, as it is sketched in fig. 2. Results obtained cover the sensitive energy range from 0 to 1.5 MeV.

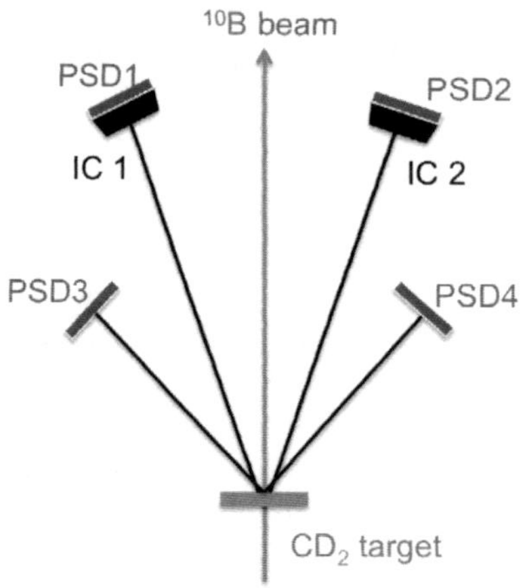

Figure 1. Scheme of the experimental setup used in the scattering chamber

^{7}Li and α were the outcoming particles detected with the use of four Position Sensitive Detectors (PSD), symmetrically placed inside the scattering chamber, setted as in fig. 1.

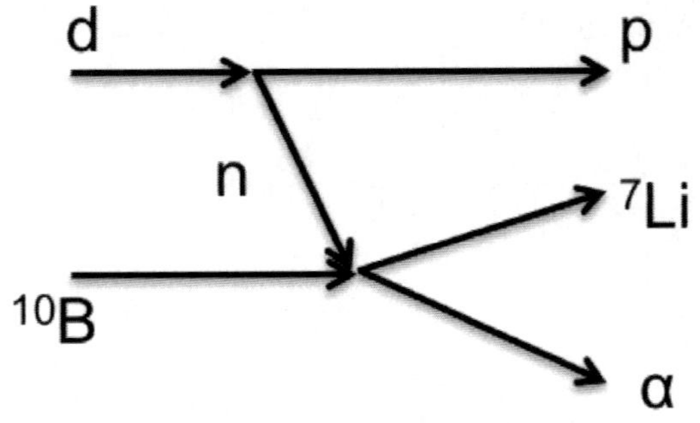

Figure 2. Sketch of the transfer reaction measured at LNS. The deuteron inside the CD_2 target has been the neutron virtual source, while the ^{10}B was the beam. ^{7}Li and α particles have been detected, while the spectator proton kinematics has been reconstructed.

The goal of separating the two contributions was achieved using a very thin target (56 μg/cm^2), minimizing the energy straggling and loss of the beam and the outcoming particles.

Once α and ^{7}Li produced have been discerned via their energy loss in isobutane gas inside ionization chambers placed in front of two of the four PSD in the scattering chamber, the Q-value spectrum, reported in [16], clearly shows two separated peaks related to the channels desired. Some details are reported in [16] and [17].

Angular distributions have been extracted thanks to the avoidance of the centrifugal barrier for each level foreseen by literature in the measured E_{cm} range. They result to be well in agreement with what reported in literature, as shown in fig. 3 where the case of 11.600 MeV level is plotted (α_0 data set). In this figure blue and red dots mark different kinematical conditions of the experimental run, while the black line is the theoritical distribution, calculated as [19] for the J_π foreseen by literature, namely 5/2+ [18].

4. Conclusions

The THM extension to neutron induced reactions has proved to be a very powerful instrument to overcome centrifugal barrier effects, helping in the study of levels which are suppressed in direct measurements, allowing the possibility to measure angular distributions and nuclear properties and obtaining the strenght of each level, thanks to the new Modified R-matrix approach [20].

Moreover, it perfectly suits complex situations of measurements (including RI beams case), because the use of deuteron as a virtual neutron source permits the use of simple setups, easy targets and the need of only one beam energy to get a wide energy range in the excitation function, finally making simple and less expensive the research.

[1] Schulze R. et al 2013 *Journal of Analytical Atomic Spectrometry* **28** 1508-1512
[2] Spitaleri C. 2016 *European Physical Journal A* **52** 77
[3] Tribble R. E. et al., 2014 *Reports on Progress in Physics* **77** 10
[4] Tumino A. et al. 2014 *Astrophysical Journal* **785** 2
[5] La Cognata et al. 2012 *Astrophysical Journal* **777** 2
[6] Tumino A. et al. 2005 *European Physical Journal A*
[7] Gulino M. et al. 2010 *Journal Physics G* **37**
[8] Kappeler F. et al. 1989 *Reports on Progress in Physics* **52** 8
[9] Guardo G. L. et al. 2017 *Physical Review C* **95** 025807
[10] Lamia L. et al. 2008 *Nuovo Cimento* **31** C

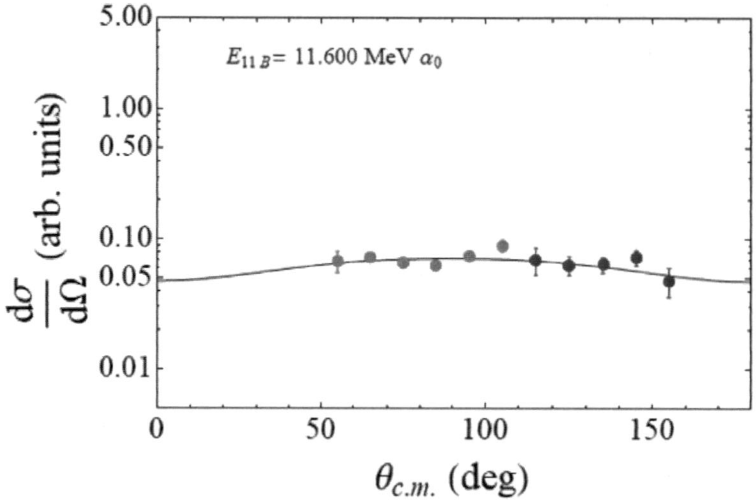

Figure 3. Angular distribution of the ^{11}B level at 11.600 MeV for the α_0 data obtained. Red and blue dots correspond to different kinematical conditions of the experiment, compared to the theoretical angular distribution (black line) calculated as in [19].

[11] Lamia L. et al. 2017 *Eueropean Physical Journal Web of Conferences* **165** 01032
[12] Hayakawa S. et al. 2018 *AIP Conference Proceedings* **1947** 020011
[13] Sergi M. L. et al. 2016 *LNS Activity Report*
[14] Barth F. F. et al. 1992 *Cancer*, 70-12
[15] Yanch J. C. et al. 1999 *Medical Physics* **26** 364
[16] Spartá R. et al. 2016 *LNS Activity Report*
[17] Spartá R. et al. 2015 *Journal of Physics: Conference Series* **703**
[18] Kelley J. H. et al. 2012 *Nuclear Physics A* **880** 88-195
[19] Blatt J. M. et al. 1952 *Review of Modern Physics* **24** 258
[20] La Cognata M. et al. 2007 *Physical Review C* **76** 065804

IOP Conf. Series: Journal of Physics: Conf. Series **1308** (2019) 012023 doi:10.1088/1742-6596/1308/1/012023

The fun (?) of rare event searches

Kai Zuber

Institut für Kern- und Teilchenphysik, TU Dresden, 01069 Dresden, Germany

E-mail: zuber@physik.tu-dresden.de

Abstract. Various kinds of rare events can be studied but only a small fraction of these are mentioned in these proceedings. The search for rare events is quite some fun but also very hard work as typically very high backgrounds must be tackled for a very small signal. The first process discussed is double beta decay, which is an extremely rare process and requires half-live measurements around 10^{20} years for the neutrino accompanied mode, while for the neutrino-less mode much longer half-lives are already explored. Nuclear physics input is needed for the matrix elements and a quenching measurement of g_A will be presented. In this case the focus is on the highly forbidden Cd-113 beta decay. New half-lives for the forbidden EC/beta decay measurement of V-50 will be presented. In addition new precision half-lives of long-living alpha decays will be presented as well.

1. Introduction

Neutrinos play a crucial role in modern particle, nuclear and astrophysics including cosmology [1, 2, 3]. It has been the major achievement of the last 20 years to show that neutrinos have a non-vanishing rest mass. The evidence arises from a deficit of upward going atmospheric muon neutrinos within the Super-Kamiokande experiment confirmed by long baseline accelerator experiments later and the solution of the solar neutrino problem by the SNO experiment being confirmed by nuclear reactor measurements. Both observations were awarded the Nobel price of physics 2015. All the mentioned observations can be explained by neutrino oscillations, which are depending on mass differences $\Delta m^2_{ij} = m^2_j - m^2_i$ with $i, j = 1, 2, 3$ as the three mass eigenstates. Furthermore, as we are dealing with a 3 $\times$3 unitary matrix, three more angles and one complex phase are involved. In case of neutrino-less double beta decay (see below) two more complex phases will appear. However, oscillation results do not provide absolute mass values. Hence, the determination of absolute neutrino masses is now a major issue. The classical way to search for a rest mass of the neutrino is the study of the endpoint region of electron spectra in beta decay (see [6] for a review). The KATRIN tritium experiment has started data taking recently. As beta decay measures the mass of $\bar{\nu}_e$ also activities have been started to use ν_e. For that the endpoint of the internal bremsstrahlung of the EC of Ho-163 is investigated by the two experiments ECHO and HOLMES. Finally, further bounds on the total sum of neutrino masses can be obtained from cosmological studies, dominated by recent Planck satellite measurements combined with large scale structure information based on baryonic acoustic oscillations (BAO) suggesting that the summed mass of all three states is less than about 0.15 eV.

IOP Conf. Series: Journal of Physics: Conf. Series **1308** (2019) 012023 doi:10.1088/1742-6596/1308/1/012023

2. Double beta decay

Another laboratory process to search for neutrino masses is the rare nuclear decay of neutrinoless double beta decay

$$(Z, A) \to (Z+2, A) + 2e^- \quad (0\nu\beta\beta\text{-decay}) \tag{1}$$

accompanied by the Standard Model process of

$$(Z, A) \to (Z+2, A) + 2e^- + 2\bar{\nu}_e \quad (2\nu\beta\beta\text{-decay }). \tag{2}$$

Single beta decay must be forbidden or at least strongly suppressed to observe this decay and only 35 potential double beta emitters exist in nature. As can be seen from Equation 1 the given decay mode is violating total lepton number by two units because the neutrino has to be its own anti-particle. If found this would be physics beyond the Standard Model. Being a decay the observable is a half-life which can be linked to the quantity of interest ϵ via

$$(T^{0\nu}_{1/2})^{-1} = G_{PS} \mid NME \mid^2 \epsilon^2 \tag{3}$$

with G_{PS} being the phase space, $\mid NME \mid$ as the involved nuclear matrix element for the physics process considered to describe this decay and ϵ the quantity of interest. Independent of the dominant decay mechanism, if $0\nu\beta\beta$-decayis ever observed experimentally it will imply that neutrinos are Majorana particles [7]. Furthermore, given all the possible processes it will become an important question what the individual contributions of the considered processes will be. For a recent review on the particle physics of double decay see [3, 8].

The standard interpretation considered is the one using light Majorana neutrino exchange. In this case ϵ is the effective Majorana neutrino mass $\langle m_{ee} \rangle$ given by

$$\epsilon \equiv \langle m_{ee} \rangle = \mid \sum_i U_{ei}^2 m_i \mid \tag{4}$$

with U_{ei}^2 as the Pontecorvo-Maki-Nakagawa-Sakata (PMNS)-mixing matrix elements [4, 5] containing the electron neutrino. The behaviour of $\langle m_{ee} \rangle$ as a function of the lightest mass eigenstate using the current oscillation results is shown in Figure 1. For $\langle m_{ee} \rangle$ values larger than about 100 meV neutrinos are almost degenerate, the inverted hierarchy covers a range between about 10-50 meV and below 10 meV is the region of the normal hierarchy. As can be seen, in the NH there is a chance for cancellation among the terms (Fig. 1). There is no such effect in the IH because of the non-maximal solar mixing angle θ_{12}. Half-lives for the IH are in the region beyond 10^{26}years while half-lives in the NH are well beyond 10^{28} years.

Recently, a better understanding of the NME became a larger issue [9], and new experimental data have been accumulated to provide better input data for the theoretical calculations. This includes - among others - charge exchange reactions, muon capture, nucleon transfer reactions and the impact of deformation of nuclei in the transition.

2.1. General experimental considerations

Evidently measurements of half-lives around 10^{20} years and well beyond are by no means trivial. The signal for the $0\nu\beta\beta$-decay(see Equation 1) is a peak at the Q-value in the sum energy spectrum of the two electrons. From equation (1) it is also apparent that the maximal information obtainable from measurements will be the single electron energies, the opening angle between them and the daughter ion. This might be accompanied by characteristic gamma rays

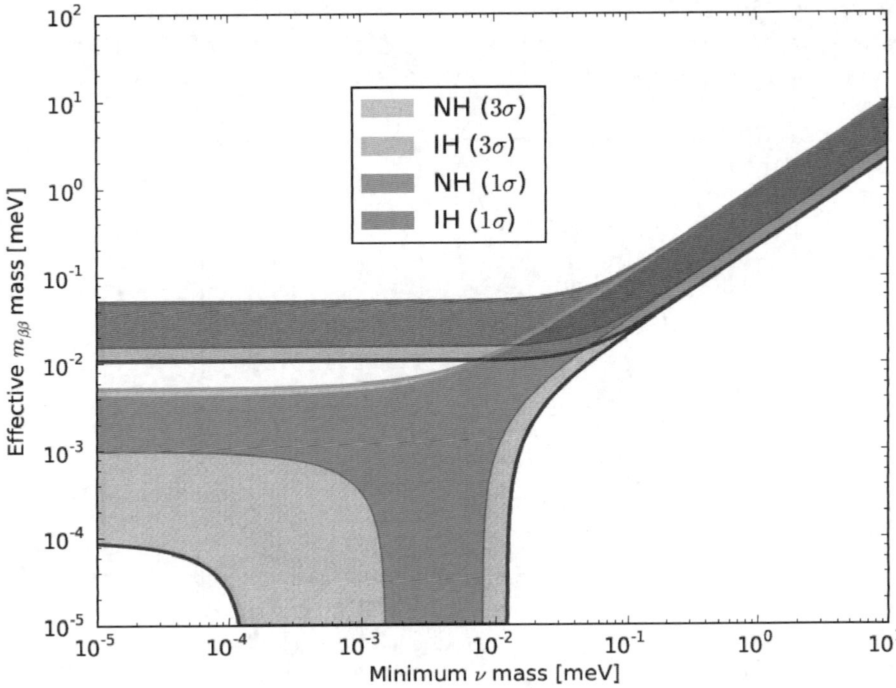

Figure 1. A presentation of the effective Majorana neutrino mass $\langle m_{ee}\rangle$ as a function of the lightest mass eigenstate. Shown are the inverted (green) and normal (red) hierarchy bands using the current values of the mixing angles and their 1σ and 3σ range. As can be seen within the normal hierarchy there can be perfect cancellation of terms in case that the lightest neutrino mass is in the range $10^{-3} - 10^{-2}$ eV (1σ) or even $10^{-4} - 10^{-2}$ eV (3σ).

in case of excited state transitions. The corresponding half-life in case of no background is given by the radioactive decay law (assuming a measuring time $t \ll T^{0\nu}_{1/2}$)

$$T^{0\nu}_{1/2} = \ln 2 m a t N_A / N_{\beta\beta} \tag{5}$$

with m the used mass, a the isotopic abundance of the double beta emitter, t the measuring time, N_A the Avogadro constant and $N_{\beta\beta}$ the number of double beta events, which has to be taken from the experiment. If no peak is observed and a constant background (in general terms these are all potential energy depositions in the region of interest, i.e. around the Q-value, not being neutrino-less double beta decay) is assumed scaling linearly with time, a half-live estimate can be derived as

$$(T^{0\nu}_{1/2})^{-1} \propto a \times \epsilon \sqrt{\frac{M \times t}{B \times \Delta E}} \tag{6}$$

where ϵ is the efficiency for detection of the total energy of both electrons, ΔE is the energy resolution at the peak position and B the background index normally given in counts/keV/kg/year. Hence, the most crucial parameters are a high detection efficiency and high abundance of the isotope of interest. This is the reason why almost all next generation experiments are using enriched materials and the "source = detector" approach, i.e. the emitter is part of the detector itself. Furthermore, the energy resolution[1] should be as good as possible to concentrate the few expected events in a small region and ideally the experiment should be background free. An irreducible background is the Standard Model process $2\nu\beta\beta$-decay . Here

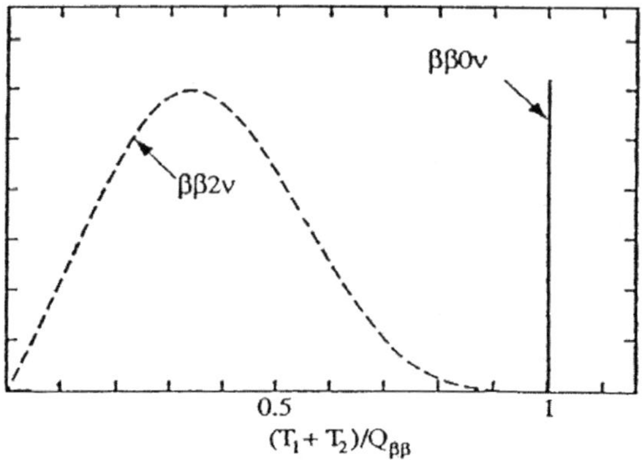

Figure 2. Schematic plot of the sum energy spectrum for the two electrons in double beta decay. While the $2\nu\beta\beta$-decay shows a continuous energy distribution the $0\nu\beta\beta$-decay b is producing a peak at the endpoint (Q-value).Various modes can be characterised by the phase space dependence $(\text{Q-E})^n$. The mode n=5 is the $2\nu\beta\beta$-decay while the modes n=1,3,7 involve the emission of a majoron, a Goldstone boson linked to the spontaneous breaking of lepton number. The different modes belong to different behaviours of the majoron with respect to weak isospin. The individual contributions are not to scale.

again energy resolution matters, because of the continuous spectrum of the $2\nu\beta\beta$-decay mode, its high energy part is leaking into the peak region. Nevertheless, this can be a worry as the half-life is typically several orders of magnitude shorter than the expected one for $0\nu\beta\beta$-decay.

2.2. Current status

As the decay rate for $0\nu\beta\beta$-decayscales with Q^5 only isotopes with Q-values above 2 MeV are considered for experimental searches. They are listed together with their natural abundance and Q-value in Table (1). The current status with of lower half-limits with at least 10^{25} years is given in Tab. 2.

As the current results point towards an upper limit of about 100 meV (given the NME uncertainties), the next goal will be to reach the region of the inverted hierarchy, i.e $\langle m_{ee} \rangle$ below ≈ 50 meV. The mentioned experiments plan upgrades (LEGEND, nEXO) and new experiments like SNO+ and others will add more data in the future.

2.3. Quenching of g_A

Recently, another quantity was brought to attention, namely the "quenching" of the axial-vector constant g_A which enters into Eq.3 with g_A^4. Recent studies showed that this quenching can span a wide range [11]. It was suggested [12] that highly forbidden beta decays like Cd-113 should have some power to measure g_A as the spectral shape of the decay electrons is changing. This measurement has been done by the COBRA collaboration [13] which is running 64 CdZnTe

[1] Care must be taken when comparing experiments, as for traditional reasons different detector technologies use either the Gaussian σ or the Full Width at Half Maximum ΔE to quote energy resolution. The relation among the quantities is $\Delta E = 2.35\sigma$.

Isotope	nat. abund. (%)	Q-value (keV)
^{48}Ca	0.187	4262.96 ± 0.84
^{76}Ge	7.8	2039.006 ± 0.050
^{82}Se	9.2	2997.9 ± 0.3
^{96}Zr	2.8	3356.097 ± 0.086
^{100}Mo	9.6	3034.40 ± 0.17
^{110}Pd	11.72	2017.85 ± 0.64
^{116}Cd	7.5	2813.50 ± 0.13
^{124}Sn	5.64	2292.64 ± 0.39
^{130}Te	34.5	2527.518 ± 0.013
^{136}Xe	8.9	2457.83 ± 0.37
^{150}Nd	5.6	3371.38 ± 0.20

Table 1. Table showing the eleven candidate isotopes with a Q-value larger than 2 MeV. Given are the natural abundances and Q-values as determined from precise Penning trap measurements or from the Atomic Mass Evaluation 2016 [10].

Isotope	$T_{1/2}^{0\nu}$	Exp.
^{136}Xe	10.7	KL-Zen
^{76}Ge	8	GERDA
^{76}Ge	1.8	MAJORANA
^{136}Xe	1.8	EXO-200
^{130}Te	1.5	CUORE

Table 2. Experimental lower half-live limits above 10^{25} years (as of end 2018) for the $0\nu\beta\beta$-decayin units of 10^{25} yrs. Also given are the name of the experiments and the selected isotope.

semiconductor detectors in the Gran Sasso Underground Laboratory (LNGS). Various templates of three different theoretical models have been studied and point towards (at least for 2 models) a value of about 0.92 [14] (Fig. 3). Hence, the value of quenching has quite some impact on the mass determination.

3. Alternative processes including positrons and electron capture

An equivalent process to the one discussed is $\beta^+\beta^+$-decay also in combination with electron capture (EC). There are three different variants possible depending on the Q-value:

$$(Z,A) \rightarrow (Z-2,A) + 2e^+(+2\nu_e) \quad (\beta^+\beta^+) \tag{7}$$
$$e_B^- + (Z,A) \rightarrow (Z-2,A) + e^+(+2\nu_e) \quad (\beta^+/EC) \tag{8}$$
$$2e_B^- + (Z,A) \rightarrow (Z-2,A)(+2\nu_e) \quad (EC/EC) \tag{9}$$

$\beta^+\beta^+$ is always accompanied by EC/EC or β^+/EC-decay. The positron production reduces the effective Q-value by $2m_ec^2$ per positron. Therefore, the rate for $\beta^+\beta^+$ is small and energetically only possible for six nuclides, however it would have a striking signature with four 511 keV gamma rays. It was shown that the β^+/EC-mode has an enhanced sensitivity to right-handed weak currents [15] and might be valuable to explore if $0\nu\beta\beta$-decayis discovered. The full Q-value is available in the EC/EC mode which is the hardest to detect experimentally. However, it was proposed [16, 17] that if an excited state of the daughter nucleus is degenerate

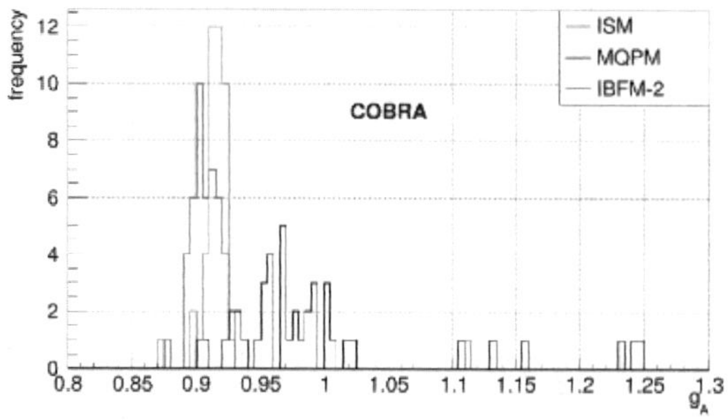

Figure 3. A plot of all best fit-values of g_A as obtained from the measured Cd-113 spectra with the COBRA CdZnTe semiconductor detectors using various theoretical templates of different g_A. As each detector has enough statistics a histogram could be done. The three nuclear models used are the interacting shell model ISM (blue), the microscopic quasiparticle-phonon model (MQPM,red) and the Interacting boson fermion model (IBFM-2,black).

with the original ground state a resonance enhancement in the decay rate could occur and the de-excitation gammas would serve as a nice signal. Due to the sharpness of the resonance a more detailed study of candidates had to wait for Penning traps entering the field and exploring reasonable candidates. The most reliable one seems to be ^{152}Gd (see Figure 4) where such a scenario is realised [18]. Despite this nice effect, to achieve the same sensitivity of $\langle m_{ee} \rangle$ as in $0\nu\beta\beta$-decayseems to require a measurement of the half-life an order of magnitude longer making this method slightly less attractive. Decays have been searched for in a number of isotopes with half-life limits ranging from $10^{15} - 10^{21}$ years, but no signal could be found in direct searches.

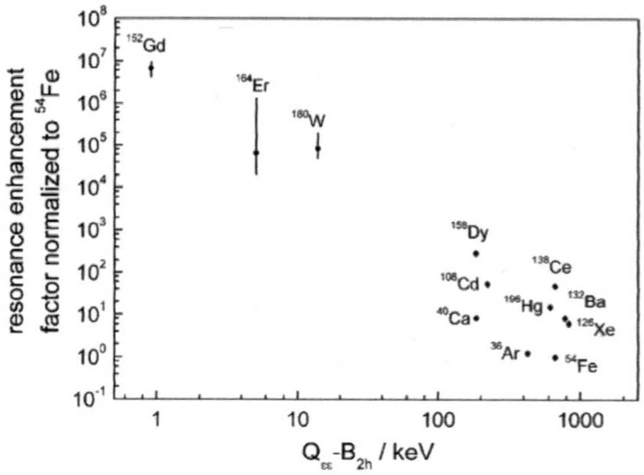

Figure 4. Compilation of enhancement factors in resonant neutrino-less double electron capture based on mass measurements with Penning traps (from [18]).

4. Other exemplaric measurements with new results of long-living decays

4.1. The decay of V-50

As Cd-113 has already been mentioned as a highly forbidden non-unique decay, another highly forbidden beta decay is V-50. Its decay scheme is shown in Fig. 5. A first measurement has

IOP Conf. Series: Journal of Physics: Conf. Series **1308** (2019) 012023 doi:10.1088/1742-6596/1308/1/012023

already been performed about 70 years ago and over time upper limits and positive observations are going hand in hand. A very clear first EC half-life measurement is reported in [19]. Theoretical calculations have been performed to predict the potential beta decay channel [20]. In a new, improved measurement using the LNGS ULB HPGe detector the EC-half life has been confirmed and the uncertainties reduced as well as an improvement by one order of magnitude in the beta decay channel [21]. The current values are now

$$\begin{aligned} T_{1/2} &> 1.9 \times 10^{19}\text{yrs } (90\% C.I.) \quad \text{beta decay} \\ T_{1/2} &= 2.67^{+0.16}_{-0.18} \times 10^{17}\text{yrs } (68\% C.I.) \quad \text{EC decay} \end{aligned} \tag{10}$$

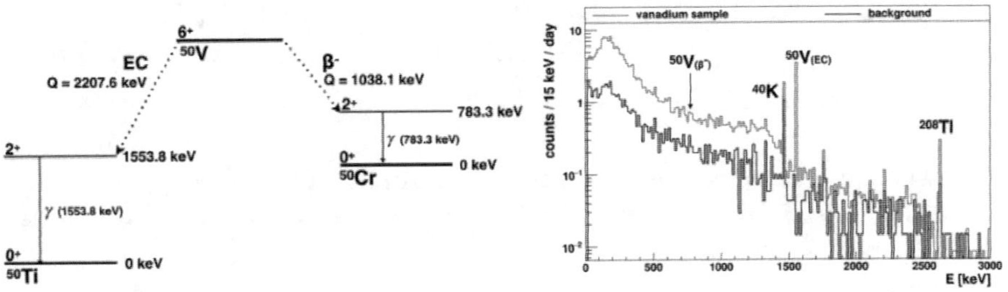

Figure 5. Left: Decay scheme of V-50, showing the EC and beta branches ($6^+ \to 2^+$) and the corresponding γ energies. Right: Measured spectrum of a Vanadium sample for 240 days at the ULB HPGe detectors at LNGS. Clearly visible is the EC-line. Frustratingly, the Compton scattering of the EC-line causes a major background for a search of the V-50 beta decay line [21].

4.2. The α decay of Sm-147 and Pt-190

With a newly built low background Frisch grid ionisation chamber [22] long living $\alpha-$decays can be explored. According to the Geiger-Nutall rule the region between 1-4 MeV is covering a variety of alpha emitters between 10^{6-17} years. The background in the chamber in this region is only 1 event per 2 days. In a first measurement a comparison with the Sm-147 standard has been performed with very high statistics [23] and agrees very well with previous measurements, our precision is better than most of the Sm-measurements. After that Pt-190 has been measured which has a link to geo-chronometers. The abundance is only 0.02% The according peak(s) can be seen in 6. The alpha-peaks are clearly well beyond the background. This measurement is the first ever which agrees with geochronological studies.

The established precise half-life measurements are [23, 24]

$$\begin{aligned} T_{1/2} &= 1.097(26) \times 10^{11}\text{years} \quad (\text{Sm-147}) \\ T_{1/2} &= (4.97 \pm 0.16) \times 10^{11}\text{years} \quad (\text{Pt-190}) \end{aligned} \tag{11}$$

5. Summary

As much it is fun to search for rare events and very long-living isotopes it is also very hard work. In this proceeding the major issue was on neutrino-less double beta decay where now half-life limits around 10^{26} years are studied. A study of highly forbidden beta decay of Cd-113 showed that the so called "quenching of the axial-vector constant" indeed occurs as the

IOP Conf. Series: Journal of Physics: Conf. Series **1308** (2019) 012023 doi:10.1088/1742-6596/1308/1/012023

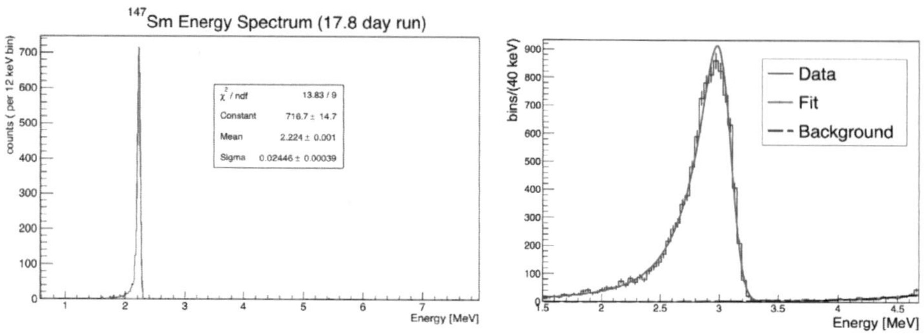

Figure 6. Left: The Sm-147 decay peak including background. The plot is showing 17 days of data from a longer measuring campaign. Right: A new measurement of Pt-190 of about 79 days. Notice the change in energy range in both plots, which is blown up for Pt-190.

measured value is significantly different from the normal value. Finally a decades long study of the V-50 half-life was improved again and touching theoretical calculations. Last but not least, also long-living α-decays were presented, with improved precision which also adding to cosmo-chronometer dating.

[1] K. Zuber, *Neutrino physics*, CRC press 2012
[2] W. Rodejohann, *Int. Journal of Modern Physics* **20**, 1833 (2011)
[3] H. Ejiri, J. Suhonen, K. Zuber, *Phys. Rep.* **797**, 1 (2019)
[4] B. Pontecorvo, *Sov. Journal JETP* **7**, 172 (1958)
[5] Z. Maki, M. Nakagawa and S Sakata, *Prog. Theo. Phys.* **28**, 870 (1962)
[6] E.Otten, C. Weinheimer, *Rep. Prog. Phys.* **71**, 086201 (2008)
[7] J. Schechter, J.W.F. Valle, *Phys. Rev.* **D 23**, 1666 (1981)
[8] M. J. Dolinski, A.W.P Poon, W. Rodejohann, arXiv:1902.04097, to published in Ann. Rev. Nucl. Part. Sci.
[9] K. Zuber, nucl ex-0511009
[10] M. Wang et al., CPC **41**, 1 (2016)
[11] F. F. Deppisch, J. Suhonen, *Phys. Rev.* C **94**, 055501 (2016)
[12] J. Kostensalo, M. Haaranen, J. Suhonen, *Phys. Rev.* C **95**, 044313 (2017)
[13] K. Zuber, *Phys. Lett.* **B 519**, 1 (2001)
[14] L. Bodenstein-Dresler et al., arXiv:1806.02254
[15] M. Hirsch et al., *Z. Phys.* A **347**, 151 (1994)
[16] J. Bernabeu, A. DeRujula, C. Jarlskog, *Nucl. Phys.* **B 223**, 15 (1983)
[17] Z. Sujkowski, S. Wycech, *Phys. Rev.* C **70**, 052501 (2004)
[18] S. Eliseev et al., *Phys. Rev. Lett.* **106**, 052504 (2011)
[19] H. Dombrowski, S. Neumaier, K. Zuber, *Phys. Rev.* C **83**, 054322 (2011)
[20] M. Haraanen, P. C. Srivastava, J. Suhonen, K. Zuber, *Phys. Rev.* C **90**, 044314 (2014)
[21] M. Laubenstein, B. Lehnert, S. S. Nagorny, S. Nisi, K. Zuber, *Phys. Rev.* C **99**, 045501 (2019)
[22] A. Hartmann et al., *Nucl. Instrum. Methods* A **814**, 12 (2016)
[23] H. Wilsenach et al., *Phys. Rev.* C **75**, 034618 (2017)
[24] M. Braun et al., *Phys. Lett.* **B 768**, 317 (2017)

AUTHOR INDEX

AUTHOR INDEX

AUTHOR INDEX